Thomas R. Willemain

Statistical Methods for Planners

The MIT Press
Cambridge, Massachusetts, and London, England

Second printing, 1981
© 1980 by
The Massachusetts Institute of Technology

This book was set in Times New Roman by Asco Trade Typesetting Ltd, Hong Kong, and printed and bound by Halliday Lithograph in the United States of America.

Library of Congress Cataloging in Publication Data

Willemain, Thomas R
  Statistical methods for planners.

  Includes bibliographies and index.
  1. City planning—Statistical methods. I. Title.
HT166.W533   307.7'6'028   79-27569
ISBN 0-262-23101-8

To Lucinda Rae

Contents

Appendixes

Preface

This text is designed to form the core of a one- or two-semester introductory course for planning students at the graduate or advanced undergraduate level. It proceeds from several beliefs about planning education. First, I believe that planning students have an acute need to be able to document and process the uncertainty involved in three common planning tasks: making estimates, making predictions, and making comparisons. There is usually a high degree of ambiguity in both the normative and empirical realms of planning; professional excellence requires that planners confront this ambiguity with competence. Part of that competence involves the ability to assess empirical uncertainty. This is especially true as the work of planners becomes more public and formal through their participation in regulatory and quasi-regulatory activities, in which the standards of evidence are more stringent than in traditional plan development. The importance of statistical competence is underscored by the recent proliferation of required courses in planning and other professional schools throughout the United States.

Second, I believe that it is possible and proper for an introductory course to develop in students having limited mathematical preparation an ability to execute modest-sized statistical analyses. When designing a course, we should keep in mind the condition of the students several years hence. A course that aims low, seeking to train "educated consumers," is likely to produce only reluctant consumers who in time lose their shallow competences. At today's tuition prices that is not an impressive legacy. A course that aims higher and routinely takes the students to the point of doing rather than talking about analyses will leave students better off no matter how much further statistical work they undertake. Beware the student who approaches the subject vowing that he can always hire some consultant to do the statistics for him—such students show more than one kind of naiveté about the use of consultants and small appreciation for a marketable skill.

Despite the good reasons for acquiring statistical competence through a course that asks students to be active, courses about methods for planners are very often troubling for students, especially when required. My third belief is that some significant part of this problem arises because of the paucity of good teaching materials. This text was developed to help fill the need for a stimulating treatment of statistics that uses the idiom of the planner. It should be easy to empathize with the student dragooned into attending a class that drones on about red and black balls inside urns, or trudges from formula to formula. Statistics can be a planner's craft, and so can be about problems of substantive interest to planning students. Several excellent books inspired this one but were not written with planning students in mind. My debt to their authors will be clear to anyone who looks at the supplementary readings. My aim is to interpret their ideas for planners.

This text assumes a limited background in mathematics. Operationally, this translates into a working knowledge of algebra, graphing of functions, logarithms and exponentials, and the $\sum$ notation. No prior knowledge of probability or statistics is assumed beyond a nodding acquaintance with tables and histograms. Calculus does not appear. The text does presume certain equipment: a supply of graph paper (linear, semi-log, log-log, and probability) and a hand calculator (preferably programmable). During the years I taught the course at MIT, there were weekly computer exercises as well, and I recommend at least a few of these as supplements. By all means students should have access to standard census data, as in the *County and City Data Book*. Every student has theories and should be encouraged to test them empirically (for too many, this will represent their first confrontation of private theory and public data).

Some important topics not included in the text which I emphasized in my course at MIT are making data (survey research, participant observation techniques, and creation of indices) and making decisions (decision analysis and linear programming). Nor are spatial probability and time-series techniques (like exponential smoothing and cohort-survival methods) included. To limit the size and sharpen the focus of the text, it proved necessary to restrict the range of topics to the statistical aspects of making estimates, making predictions, and making comparisons.

Several features of the text should make it more effective for planners. Perhaps the major feature is the strong, but not exclusive, focus on Bayesian methods of estimation. Bayesian methods are especially appropriate for planners for two reasons. First, the likelihood function is a more natural basis for inference than the null hypothesis: planning students instinctively know that sample results are more or less consistent with many possible values of a parameter, whereas the traditional null hypothesis rightly appears to be somewhat stilted and artificial. Second, planners are highly invested in their sense of professional judgment, and the prior distribution offers them a graceful and constructive outlet for that professional judgment. My sense is that Bayesian methods are inherently more teachable than conventional approaches. Instructors who are used to confidence intervals will find the Bayesian construct of highest density regions numerically similar (if conceptually somewhat different). When Bayesian approaches are not available or too complicated, traditional significance testing material is presented instead.

Several other features should be noted. First, illustrations within each chapter are drawn from areas of substantive interest to planners. This facilitates the transition from purely statistical concerns to the kind of substantive discussion that ultimately makes for the most meaningful statistical analysis. Second, each chapter contains a few problems chosen the same way. Typically, a problem will contain a small data set and ask for an analysis; often the questions will lead the student somewhat beyond the text; frequently there is no unique solution to the problem. There are few straightforward computational problems—these should be prescribed as needed and should definitely be based on data sets of personal interest to the student. Third, each chapter contains references and supplemental readings. Often the readings are listed because they contain fascinating data, controversial findings, or useful paths into the deeper reaches of a topic. Many of the readings have proved themselves as good bases for classroom discussions; some are meant for only the most advanced students. Fourth, at several points the text stresses the illuminating use of synthetic data and Monte Carlo simulation as vehicles for nurturing a feel for randomness. Fifth, stress is laid on the importance of the assumptions underlying inferential statistics, on simple graphical methods to test the assumptions informally, and on how to react to apparent violations of the assumptions (this is especially important for those who will work in judicial or quasi-judicial environments). Sixth, some of the most useful recent writings on the problems of executing and interpreting regression analyses have been interpreted for the student in a first course. Included are such topics as outliers, multicollinearity, errors in measurement of variables, and making causal interpretations. Seventh, the issue of the internal validity of experimental designs is illuminated through use of a simple mathematical model of the nonequivalent control group design and its constituent parts.

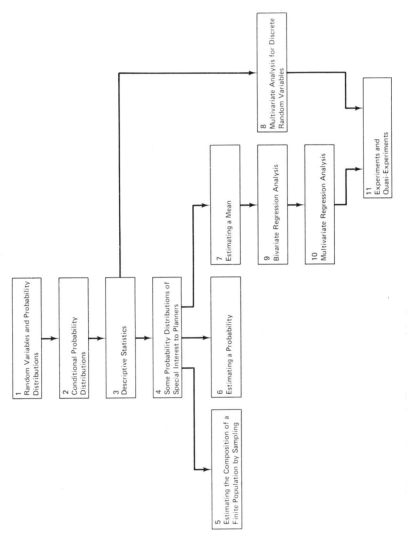

Flow of chapters (arrows indicate prerequisites)

The flow of the material is illustrated in a diagram which shows the prerequisites linking chapters. Chapter 1 begins with a brief introduction to probability and random variables from the perspective of making primitive predictions. Chapter 2 introduces conditional probability from the perspective of refining predictions and develops Bayes' rule for later use in estimation. Chapter 3 treats descriptive statistics as portable summaries of probability distributions, stressing the relationship between choice of descriptive statistic and level of measurement. Chapter 4 describes five probability distributions of special interest to planners.

Chapters 5 through 7 present Bayesian methods for estimation. Chapter 5 addresses the problem of estimating from a sample the number of each of two kinds of cases in a population of finite size. Chapter 6 takes up the estimation of a probability. Chapter 7 concerns estimating the mean value of a variable in a large population from a small sample.

Chapters 8 through 10 focus on using attributes of a case to refine predictions. Chapter 8 deals with crosstabulations, presenting measures of strength of association and the chi-square test for the statistical significance of an association; the chapter also introduces the issues of multicollinearity and causal inference. Chapters 9 and 10 deal with bivariate and multivariate regression, respectively. Together the two chapters discuss data transformations, uncertainty in prediction, dummy variables, statistical inference, and problems in interpreting regression coefficients.

Chapter 11 focuses on comparative evaluation and experimental design, stressing the link between the structure of comparisons and their vulnerability to threats to internal and external validity. Chapter 11 also discusses randomization, matching, and the Mann-Whitney nonparametric test of the significance of differences between two sets of measurements.

Acknowledgments

Several colleagues at the MIT Department of Urban Studies and Planning helped me arrange my thoughts over the three years I spent teaching Methods for Planners. I owe a special debt to Professor Joseph Ferreira, Jr., who was ever ready to listen and suggest ideas, argue approaches, and report classroom experience with preliminary drafts. Professor Michael O'Hare converted me to the Bayesian perspective. Professors Thomas Nutt-Powell and Philip Clay taught the course with me and helped shape the substantive content of the text. Keith A. Stevenson and J. Mark Schuster contributed ably to the course as instructors, and Levi Sorrell, Allen Lee, Robert Bluhm, and Khalid Saeed helped as teaching assistants and interpreters of student concerns. Pamela Ersevim typed several versions of the manuscript with competence and good grace. Professors Langley Keyes and William Porter, Department Head and Dean of the School of Architecture and Planning, respectively, provided intellectual and financial support for the first draft. The Master of City Planning classes of 1977, 1978, and 1979 at MIT offered encouragement and feedback needed from start to finish.

Several excellent texts served to shape my approach to the material. These are mentioned so often in the references and supplementary readings that only the last names of the authors are used. The full references are as follows:

D. Campbell and J. Stanley. *Experimental and Quasi-Experimental Designs for Research.* Chicago: Rand McNally, 1963.

F. Mosteller and R. Rourke. *Sturdy Statistics.* (Reading, Mass.: Addison-Wesley, 1973.

F. Mosteller and J. Tukey. *Data Analysis and Regression: A Second Course in Statistics.* Reading, Mass.: Addison-Wesley, 1977.

S. Schmitt. *Measuring Uncertainty: An Elementary Introduction to Bayesian Statistics.* Reading, Mass.: Addison-Wesley, 1969.

E. Tufte, *Data Analysis for Politics and Policy.* Englewood Cliffs, N.J.: Prentice-Hall, 1974.

J. Tukey. *Exploratory Data Analysis.* Reading, Mass.: Addison-Wesley, 1977.

Finally, I acknowledge the help of my family: Lucinda, who carried me through the course as well as the book; Matthew, who made the finishing easy; and Katy, who helped me understand about connecting the dots and other methodological techniques of great power.

I USEFUL BACKGROUND

1 Random Variables and Probability Distributions: Formalizing Uncertainty

Uncertainty is everywhere. Being able to organize and process uncertainty helps distinguish good from mediocre planners. An essential planning task is making predictions in the face of uncertainty: How many children of school age will live in a new development? What peak level of traffic can be expected in a vacation resort? How long will a tenant occupy an apartment? How many graduates of a job-training program will be employed six months after graduation? How likely is it that a small town will employ a professional city manager if it grows to 50,000 people? What is the likelihood that a patient will have to wait more than an hour to be treated in a city hospital emergency room? The list of questions is endless, and the questions lurk in every corner of city planning. This chapter is about a language for answering such questions.

The language is *probability theory*. You will see only the bare bones of the theory in this book, but you should learn enough to acquire useful skills and to be able to learn more as necessary. Probability is about *random variables*, which are variables whose values are not certain ahead of time. The answer to each question above is a random variable. For instance, the number of children in the new development could range anywhere from zero to a large number. In two identical developments the numbers of children are likely to be different, and in any one development the number will change over time. If we could only peek ahead in time, we could actually count the children of the future tenants; but we cannot, so we must make an educated guess. Our education for guessing derives from our knowledge of other similar developments and of typical tenants and perhaps from our own intuitions about the particular development in question.

We express our knowledge about the possible values of a random variable in terms of its *probability distribution*. To every possible value of the random variable there corresponds a number that represents the probability the random variable will take on precisely that value. The collection of possible values and their probabilities is the probability distribution. The probability assigned to each value of the random variable is a fraction between 0 and 1.0. If the probability is zero, the corresponding value can never occur; if the probability is unity, the corresponding value will certainly occur. The full set of possible values of the random variable must be "mutually exclusive and collectively exhaustive," meaning that one and only one of the possible values will actually occur. Since the random variable must take on some value, the probabilities assigned to all the possible values must sum to 1.0.

As an example, consider the eleven towns in Massachusetts with 1975 populations of 30 to 35,000. Suppose we wish to study their type of local government. The random variable of interest is governmental structure. It can take on three values: open town meeting, representative town meeting, and city council. Note that the random variable need not be a number; in this case it is a category. Note also that the governmental structure is not random in the sense that the type of town government changes from day to day like the weather. Each town among the eleven has one form of government and has doubtless had that particular form for a long time, and if someone named a particular town and asked us its form of government, we would look up the answer. But if the question were put more generally, we would have to use a probability distribution: "What is the chance that any given town has a representative town meeting?" To answer this, we turn to the probability distribution of the random variable shown in table 1.1. The answer is that, not knowing the identity of the town (other than that it is in Massachusetts and has 30 to 35,000 people), there is a 0.37 probability that the town uses a representative town meeting. Our formal notation for expressing this result will be

Prob [town has representative town meeting] = 0.37.

Of course, any particular town either does or does not have such a

Table 1.1
Form of local government in 11 Massachusetts towns of 30 to 35,000 population

Form of government	Number of towns	Fraction of towns
Open town meeting	2	0.18
Representative town meeting	4	0.37
City council	5	0.45
	11	1.00

Source: Massachusetts League of Cities and Towns, *Municipal Directory*, 1975–1976.

forum, but all we can do when considering an anonymous town is give the relative likelihoods of the alternative forms of government.

We can now make two observations about predictions using our probability distribution. First, if someone asked us, "Which type of government do you think a given town has," we should answer, "City council." This is the modal, or most frequently occuring category, and is our best bet for a guess. We stand a 45 percent chance of being right, compared to only 37 and 18 percent for the other possible guesses. We still have uncertainty, but we have marshalled our evidence and used it to improve on a blind guess which treats all three answers as equally likely. Second, if someone asked us how many of 100 towns of 30 to 35,000 population use either form of town meeting, we have our probability distribution as a starting place. If all the towns were like those in Massachusetts, we would expect about 55 to have town meetings; if all 100 towns were in the midwest, we might expect that 55 would be an upper bound on the number, presuming that town meetings are more common in New England than elsewhere. In this case we are using the data to inform a subjective estimate of probabilities.

It is important that you be able to read and write probability distributions. There are two varieties, corresponding to discrete and continuous random variables. A *discrete random variable* only takes on values from a distinct set: the three types of local government in the example above, the number of children in the new development, the number of users of a neighborhood health center. Probability distributions for discrete random variables all look generally like that shown in figure 1.1a. The height of each vertical line represents the probability that the random variable takes on the value in question. If you stack the lines end to end they must form a line of length 1.0 (the probabilities must sum to 1.0). Note that some probabilities (like the fifth value in the graph) will be zero, meaning that the random variable can never take that value.

Continuous random variables are not so restricted in the values they can take on—there are an infinite number of possible values. An example would be the length of time spent waiting for treatment in a city hospital emergency room. There is no reason to expect this wait to occur exactly in intervals of 5 minutes or 1 minute; although we may ultimately be forced to record the time in discrete units of 1 second intervals, in principle

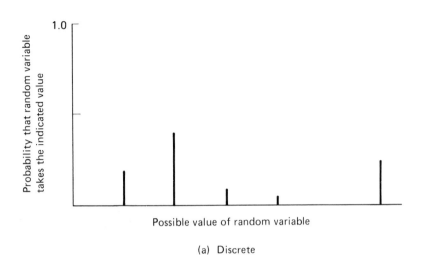

(a) Discrete

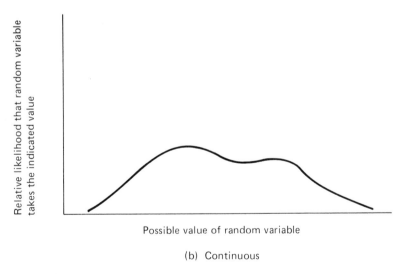

(b) Continuous

Figure 1.1
Examples of probability distributions for discrete and continuous random variables

we can treat the waiting time as if it were perfectly continuous. In the case of continuous random variables, the probability distribution is a smooth curve as in figure 1.1b, not a sequence of vertical lines, and shows not the actual probability that the random variable takes on the particular value, but the probability relative to other possible values. While the curve cannot drop below the horizontal axis, it need not stay below unity since the height of the curve represents the relative likelihood of the value below it, not the actual probability as in the discrete case. However, just as the sum of the discrete probabilities must equal unity so must the area under the curve of the continuous probability distribution. For continuous random variables we can ask only interval questions, for example, "What is the probability that the waiting time is *between* 15 and 20 minutes"; or "What is the chance that the waiting

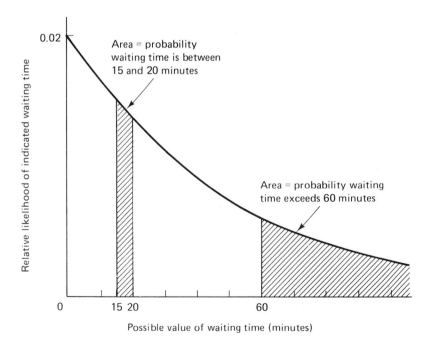

Figure 1.2
Hypothetical probability distribution of waiting time in a city hospital emergency department

time *exceeds* 60 minutes?" The answers to these questions are the corresponding areas under the curve, as shown in figure 1.2 (the entire area must equal 1.0).

Of course, we can ask similar interval questions of discrete random variables. Consider the discrete distribution shown in figure 1.3 for the number of patients who arrive at the emergency room during a certain hour. From figure 1.3 it appears that the probability that seven or more patients will arrive is less than 0.01, the probability that zero or one will arrive is about 0.40, and the probability of three to five arrivals is also about 0.40. These compound probabilities are obtained by adding together the appropriate individual probabilities: the probability of zero or one arrivals is the probability of zero arrivals plus the probability of one arrival.

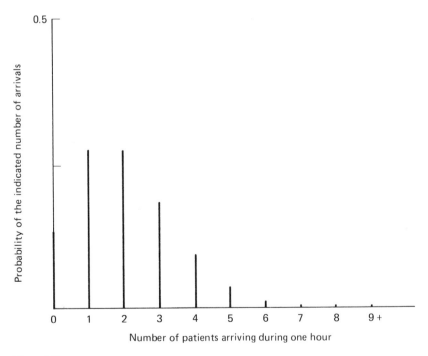

Figure 1.3
Hypothetical probability distribution of the number of patients arriving at a city hospital emergency department during one hour

Where do the probability distributions come from? That in figure 1.3 follows a theoretical pattern (a Poisson distribution with a parameter 2.0, see chapter 4), and the probabilities are tabulated in statistical reference books or can be computed from a formula. In practice, we might not have a fully developed theory, so we might rely heavily on measurements, counting the number of arrivals at the emergency room during the same hour on a number of different days, and construct a *histogram* of the number of days for which the arrivals totaled zero, one, two, and so on. Such a histogram is shown in figure 1.4, which displays the results of 26 $(2 + 7 + 9 + 3 + 4 + 0 + 1)$ observation days. On 3 of the 26 days there were exactly 3 patients arriving during the hour chosen for study; on 9 of the 26 days exactly 2 patients arrived. It is a trivial matter to convert the histogram into a legitimate probability

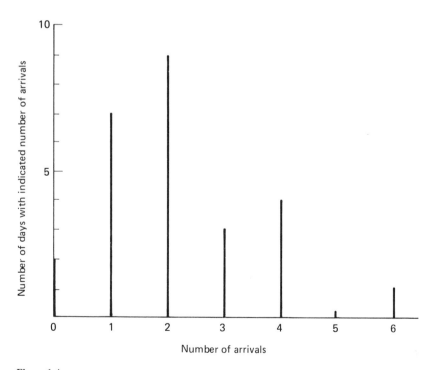

Figure 1.4
Histogram of the number of patients arriving at a city hospital emergency department during one hour

distribution: just divide each number of days by the total (26) to make the corresponding probability. Thus, for instance, the estimate from the 26 days of observations is that there is a $4/26 = 0.15$ probability of exactly 4 arrivals.

You should never fall into the trap of believing that doing statistics is a mindless, automatic process free of substantive judgment; the design of histograms offers a simple but telling instance. I use the word "design" consciously, since a statistical construct like a histogram is as much an artifact as a built structure—having many possible forms, of which one is chosen to fulfill a function. Both the component parts of the histogram and their organization must be selected carefully by the analyst. In the case of the histogram of the number of emergency department arrivals during one hour, the first design choice involves the definition of "arrival." Since some nonemergency cases arrive at the emergency department and are immediately directed to some other part of the hospital (or out of the hospital altogether), there must a decision made as to whether those people who receive no care count as arrivals. Likewise, family members accompanying a patient may count as arrivals for the facility designer who must arrange for their seating but not for the hospital administrator who must determine the physician staffing level. A second design choice involves which time periods to use for the hourly counts of arrivals. If the arrival rate varies significantly by time of day (as it almost always does for urban emergency services), then only counts for the same hour of the day can be used. But if the arrival rate also varies significantly by day of the week, then it may be necessary to gather data only once each week rather than once every day. These decisions about data pooling depend both on the nature of the random process generating the data and the use to which the histogram will be put. A third design choice involves the selection of categories for display of the data. Here the histogram designer feels two opposing pressures. To preserve detail, the designer wishes a large number of categories, yet to preserve compactness and smoothness in the display he wishes a small number of categories. This tension was resolved in figure 1.3 by combining all large numbers of arrivals into the category $9+$. The resolution is usually less obvious when the random variable in question is continuous and has no natural categories, as in the case of waiting times in the emergency department. Finally, it may happen that there are no data available on which to base a histogram, or the available data are not directly applicable (perhaps arising from the wrong time or place), so the distribution becomes an exposition of the planner's best subjective judgment. In all cases, the histogram is the planner's creation, formalizing his uncertainty.

Summary

Much of planning involves a confrontation with uncertainty—predicting the values to be taken on by random variables. By a combination of empirical and / or subjective methods we summarize in a probability distribution our knowledge of the relative likelihoods of the various possible values of a random variable. The form of the distribution varies depending on whether the random variable is discrete or continuous, but the knowledge provided is always the probability that the random variable takes on some particular value or set of values.

References and Readings

Davis, K. "World Urbanization 1950–70." In L. S. Bourne and J. W. Simmons, ed., *Systems of Cities: Readings on Structure, Growth and Policy.* New York: Oxford University Press, 1978.

Tukey. "Scratching Down Numbers," chapter 1.

Problems

1.1

The table that follows reports on ambulance response time (the time delay between calling an ambulance and its arrival at the scene of an emergency) in rural areas around Wheeling, W. Va. Use the table to plot a histogram of response time. Note that the table itself reports cumulative response time.

Time (minutes)	Percentage of calls answered
5	24
10	62
15	81
20	89
25	93
30	96
> 30	100

1.2

The following data show the number of new housing units authorized by building permits in Belmont, Mass., from 1960 to 1974. Prepare a histogram of the number of units authorized each year.

Year	Number of units authorized
1960	55
1961	67
1962	108
1963	81
1964	66
1965	37
1966	44
1967	14
1968	104
1969	12
1970	364
1971	25
1972	30
1973	27
1974	20

1.3

The following data show the number of home sales in Boston neighborhoods over a 2-year period. Use these data to make a histogram illustrating the change in level of sales from 1975–76 to 1976–77. Justify the choices you make regarding the issues of representing the changes in each neighborhood (absolute vs. percentage differences) and aggregating the cases into groups.

Neighborhood	Home sales	
	7/75 to 6/76	7/76 to 6/77
Roxbury	266	148
North Dorchester	663	479
South End	180	144
Jamaica Plain	258	200
South Boston	250	176
West End	134	84
South Dorchester	686	536
Charlestown	154	83
East Boston	275	200
Roslindale	292	218
Back Bay–Fenway	361	325
Hyde Park	300	318
Allston–Brighton	297	241
North End	45	53
West Rosbury	292	233

2 Conditional Probability Distributions: Using Additional Information to Refine Predictions

Much of the professional expertise of a planner is contained in *conditional* information. When asked the likelihood that a homeowner will sell his house within a year, the planner might simply cite the percentage of homeowners nationally who sell their homes in any year, but more likely he will press for additional information: What city? What neighborhood? What age and race of homeowner? How long has the owner lived in the house? The planner does this because these other bits of information are powerful—they can greatly change the estimate of the probability that the homeowner will sell his house. We will call these other bits of information *attributes* of the case.

In analyzing data, we strive to identify a manageably small number of attributes that can be used to reduce our uncertainty about the value some random variable will assume. Our success will depend strongly on the nature of the phenomenon of interest. We can predict rather well the number of schoolage children in a new housing development by using the attributes "number of bedrooms per dwelling unit" and "type of dwelling unit." On the other hand, we have a good deal of trouble predicting the daily cost of medical services for nursing home patients, even using many attributes: their age, sex, type of nursing home, diagnoses, functional status, and source of funds. Generally, though, we can improve predictions by conditioning them on the attributes of the cases of interest, although the degree of uncertainty remaining after being more specific about the cases of interest can still be substantial. This chapter is about how to exploit attribute data systematically to reduce uncertainty and improve predictions.

Mathematically, conditional *probability distributions* look just like unconditional distributions: conditional on a case having a certain attribute, the random variable of interest for that case must take on some one value from among all those possible; each possible value will be paired with a probability between zero and unity; the sum of all these conditional probabilities must equal unity. We will, however, modify our notation to indicate that the probability distribution is conditional. We use the vertical bar |, read "given," to separate the value of the random variable from the conditioning information, as in

Prob [sell | East Cambridge],

which represents the probability that a homeowner sells his house given that he is living in East Cambridge.

Incidently, there is no reason why we cannot consider sell / not sell an attribute of the homeowner and reverse matters by using knowledge of whether the homeowner just sold his house to help predict which neighborhood he lived in. It would certainly be legitimate to make a comparative study of those homeowners who do and do not sell their houses during the year and to focus on such conditional probabilities as Prob [lives in East Cambridge | sells]. We note that there are many ways to crossclassify a case: we have just grouped homeowners both by whether they sell their houses and by where they live.

Crosstabulation and Conditional Probabilities

One of the most common formats for analysis of data grouped into categories is the *crosstabulation*, or *contingency table*, which can be thought of as a table of counts of cases arranged into conditional distributions. Consider the crosstabulation in table 2.1 of age with limitation in activity due to chronic illness among American men. Examining the table we note that no limitation is the modal (most frequently occurring) category for each age bracket, but that, as age increases, so does the chance of having a severe limitation of activity due to chronic illness.

What is the chance that a randomly chosen American male will have a major limitation? The column sums (also called column *marginals*) show that 5,590/100,030 = 0.06, or 6 percent of American men suffer a major chronic limitation in activity. What is the chance that a randomly

Table 2.1
Crosstabulation of age with limitation in activity due to chronic illness among American men (counts in thousands)

| Age | Activity limitation | | | | |
	None	Minor	Major	Severe	Row sum
0–16	30,796	603	600	81	32,080
17–44	35,384	1,445	1,567	556	38,952
45–64	15,260	1,112	2,135	1,913	20,420
65 +	4,315	416	1,288	2,559	8,578
Column sum	85,755	3,576	5,590	5,109	100,030

Source: National Center for Health Statistics, series 10, no. 112, p. 11.

selected man is at least 65 years of age? The row marginals show that $8,578/100,030 = 0.09$, or 9 percent are at least 65 years old. Each of these questions concerned a single attribute of an American male. Now, what is the probability that an American male will possess both attributes? The table reveals that 1,288,000 men (note that the counts are in units of 1,000 men) are both over 65 and suffering from a major limitation; thus the answer is $1,288/100,030 = 0.01$.

Now we ask a question about a conditional probability: What fraction of the men less than 17 years old have severe chronic limitations of activity? To answer, we focus on the top row of the table (age 0–16) and note that

Prob [severe | 0–16] $= 81/32,080 = 0.003$.

Suppose we ask the question in reverse: What is the chance that a man with severe limitation is less than 17 years old? To answer, we focus on the rightmost column of the table and observe

Prob [0–16 | severe] $= 81/5,109 = 0.02$.

Thus 2 percent of those with severe limitations are young, but only 0.3 percent of the young have severe limitations. When we read the table vertically, we condition on activity limitation. Reading the table horizontally conditions on age.

Bayes' Rule for Conditional Probabilities

By now you should sense that computing conditional probabilities from a crosstabulation is a simple matter. Let us become a bit more formal about the definition of conditional probability, since we will arrive at a very useful formula known as *Bayes' rule*, which will later figure prominently in our treatment of estimation from samples.

How did we calculate a conditional probability like Prob [severe | 0–16]? We isolated our attention on those men aged 0–16 (the top row of the table) and found the percentage of that row sum accounted for by the rightmost column:

Prob [severe | 0–16] $= 81/32,080$.

Put another way, we found

$$\text{Prob [severe}\,|\,0\text{--}16] = \frac{\text{Prob [severe and } 0\text{--}16]}{\text{Prob } [0\text{--}16]}.$$

The numerator of the right-hand side, Prob [severe and 0–16], is the probability that a case processes both attributes of interest. When we asked the question in reverse we found

$$\text{Prob } [0\text{--}16\,|\,\text{severe}] = \frac{\text{Prob [severe and } 0\text{--}16]}{\text{Prob [severe]}} = 81/5{,}109.$$

We can rearrange both of these expressions to get two different, but equivalent, formulas for the probability that a case possesses both attributes:

$$\begin{aligned}
\text{Prob [severe and } 0\text{--}16] &= \text{Prob [severe}\,|\,0\text{--}16] \times \text{Prob } [0\text{--}16] \\
&= \text{Prob } [0\text{--}16\,|\,\text{severe}] \times \text{Prob [severe]}.
\end{aligned}$$

It follows that the right-hand sides are equal:

$$\begin{aligned}
&\text{Prob [severe}\,|\,0\text{--}16] \times \text{Prob } [0\text{--}16] \\
&= \text{Prob } [0\text{--}16\,|\,\text{severe}] \times \text{Prob [severe]}.
\end{aligned}$$

We see that we can condition either on age or on activity limitation. This may look like a shell game, but there is a real value to this flexibility. To appreciate this, you must stop to realize the marked differences between the attributes age and activity limitation. Age is a simple, oft-collected and oft-reported attribute, whereas activity limitation is more complex and expensive to measure and is not commonly available for a local population.

If one were planning services for those with severe limitations and needed a quick estimate of the age distribution of men eligible for such services, one could apply the national age-specific rates in table 2.1 to local age data in lieu of conducting a local survey. To see how this would work, we rearrange the last expression

$$\text{Prob [age } 0\text{--}16\,|\,\text{severe}] = \frac{\text{Prob [severe}\,|\,0\text{--}16] \times \text{Prob } [0\text{--}16]}{\text{Prob [severe]}}.$$

The denominator on the right-hand side serves only one function: it normalizes the expression so that the conditional probabilities will sum to 1.0:

Prob [0–16 | severe]
Prob [17–44 | severe]
Prob [45–64 | severe]
+ Prob [65 + | severe]

─────────────────────

1.0

or equivalently

Prob [severe | 0–16] × Prob [0–16] / Prob [severe]
Prob [severe | 17–44] × Prob [17–44] / Prob [severe]
Prob [severe | 45–64] × Prob [45–64] / Prob [severe]
+ Prob [severe | 65 +] × Prob [65 +] / Prob [severe]

─────────────────────

1.0

from which it follows that

Prob [severe] = Prob [severe | 0–16] × Prob [0–16]
+ Prob [severe | 17–44] × Prob [17–44]
+ Prob [severe | 45–64] × Prob [45–64]
+ Prob [severe | 65 +] × Prob [65 +].

Thus to estimate Prob [0–16 | severe], we use the counts in table 2.1 to estimate the four probabilities of severe limitation conditional on age and require three items of information about the distribution of ages among local men (we can determine the fourth using the fact that the probabilities sum to unity). The national data apply only to the extent that the local population is "typical"; we cannot check this assumption without surveying the population, but it will serve for the initial guess. We do at least use local information on age distribution. Suppose that the local age distribution for males is

Prob [0–16] = 0.30
Prob [17–44] = 0.45
Prob [45–64] = 0.15.

Since the age probabilities sum to 1.0, it must be true that

Prob [65 +] = 1 − 0.30 − 0.45 − 0.15 = 0.10.

Now we know from table 2.1 that

Prob [severe $|$ 0–16] $= 81/32,080 = 0.0025$
Prob [severe $|$ 17–44] $= 556/38,952 = 0.0143$
Prob [severe $|$ 45–64] $= 1,913/20,420 = 0.0937$
Prob [severe $|$ 65+] $= 2,559/8,578 = 0.2983.$

Therefore

Prob [0–16 $|$ severe]

$$= \frac{(0.0025)(0.30)}{(0.0025)(0.30) + (0.0143)(0.45) + (0.0937)(0.15) + (0.2983)(0.10)}$$
$$= \frac{(0.0025)(0.30)}{0.0511} = 0.0147.$$

Likewise

$$\text{Prob [17–44} | \text{severe]} = \frac{(0.0143)(0.45)}{0.0511} = 0.1260$$

$$\text{Prob [45–64} | \text{severe]} = \frac{(0.0937)(0.15)}{0.0511} = 0.2752$$

$$\text{Prob [65+} | \text{severe]} = \frac{(0.2983)(0.10)}{0.0511} = 0.5841.$$

To summarize, we can show how the difference in age structure of the local population will produce a difference in the local client population:

Age	Local	National
0–16	0.0147	0.0159
17–44	0.1260	0.1088
45–64	0.2752	0.3744
65+	0.5841	0.5009
	1.0000	1.0000

More generally, suppose the cases of interest (cities, men with chronic activity limitation, and so on) have two attributes, called A and B. Let i denote which of I possible values is taken on by attribute A, and let j denote which of J possible values is taken on by attribute B. Then Bayes' rule is

$$\text{Prob } [A = i | B = j] = \frac{\text{Prob } [B = j | A = i] \times \text{Prob } [A = i]}{\sum_{k=1}^{I} \text{Prob } [B = j | A = k] \times \text{Prob } [A = k]},$$

where the denominator is just the sum of every numerator term formed by holding the attribute B fixed and letting the attribute A take all possible values. In this example attribute A corresponds to age bracket, while attribute B corresponds to degree of activity limitation.

Often in our work we will be nearly as interested in the relative probabilities of various attributes as in their actual values (this will be especially true when the attribute we seek to predict is a continuous rather than discrete random variable, see chapters 6 and 7). In such cases we will grant ourselves the luxury of ignoring the denominator in Bayes' rule. Its only role is to guarantee that probabilities sum to unity by properly scaling the conditional distribution. For any given value of the conditioning attribute B_j, the numerical value of the denominator never changes (its value was 0.0511 in the example, which conditioned on severity of activity limitation), so all the action in Bayes' rule occurs in the numerator. Of course, when we ignore the denominator, we no longer have an exact equality; instead we write Bayes' rule as

$$\text{Prob}\,[A_i\,|\,B_j]\;\alpha\;\text{Prob}\,[B_j\,|\,A_i] \times \text{Prob}\,[A_i],$$

where the Greek letter α (alpha) is read "is proportional to." It is a simple matter to make the probabilities legitimate if we so desire later: we merely divide each quantity $\text{Prob}\,[B_j\,|\,A_i] \times \text{Prob}\,[A_i]$ by the sum of all such quantities

$$\text{Prob}\,[A_i\,|\,B_j] = \frac{\text{Prob}\,[B_j\,|\,A_i] \times \text{Prob}\,[A_i]}{\sum_{k=1}^{I} \text{Prob}\,[B_j\,|\,A_k] \times \text{Prob}\,[A_k]}.$$

Example: Monitoring the Efficiency of a Human Services Program

To cement your understanding of Bayes' rule and illustrate a general application, consider another example. Let the population of a city be divided into four groups according to eligibility for and use of a particular human services program, as in figure 2.1. Ideally, the program would operate in such a way that the counts b and c in figure 2.1 would be zero, meaning that all eligibles would be users and all users would be eligibles. In practice, it would be most unusual for a program to operate so cleanly, and it will be prudent to compute performance measures that monitor the efficiency of program operation. There are two senses of efficiency, corresponding to two conditional probabilities.

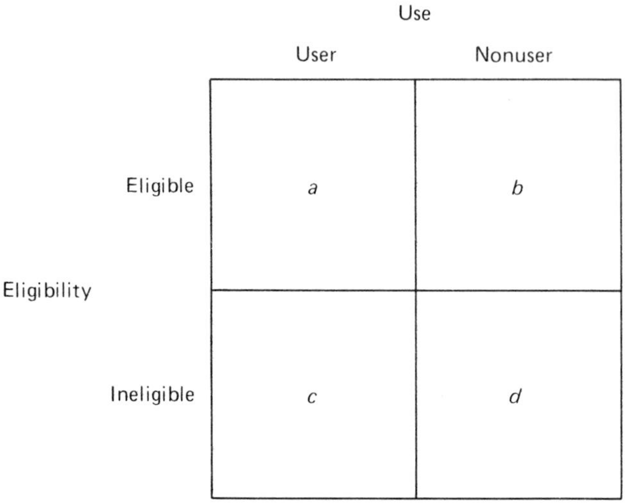

Use

	User	Nonuser
Eligible	a	b
Ineligible	c	d

Eligibility

Figure 2.1
Categorization of a city population according to eligibility for and use of a particular human services program

One concern is that eligibles make use of services. This dimension of efficiency is called *sensitivity* and is defined as

Sensitivity = Prob [use | eligible].

The other concern is that users of the service be restricted to eligible people only. This dimension of efficiency is called *specificity* and is defined as

Specificity = Prob [eligible | use].

How will sensitivity and specificity be estimated? The definitions indicate that specificity will generally be easier to estimate, since it requires information only on those individuals already in contact with the agency as clients, whereas sensitivity requires information on all eligibles, some of whom are out in the community and much less accessible to the agency. However, using Bayes' rule and the definition of specificity, we can reexpress the definition of sensitivity:

Sensitivity = Prob [use | eligible]

$$= \frac{\text{Prob [eligible | use]} \times \text{Prob [use]}}{\text{Prob [eligible]}}$$

$$= \frac{\text{Specificity} \times \text{Prob [use]}}{\text{Prob [eligible]}}.$$

Thus sensitivity can be computed by first estimating specificity from a review of clients' cases, then estimating Prob [use] and Prob [eligible]. Estimation of Prob [use] simply requires dividing the count of clients by the total population of the city. Estimation of Prob [eligible] may be relatively easy using standard census data when eligibility is determined by some simple combination of demographic factors like age, sex, and income. Other types of eligibility rules may necessitate an expensive community survey to estimate Prob [eligible], but such a needs assessment would probably have been required anyway to justify starting the program and would therefore be available already. In such a case it would be a great advantage not to have to undertake another community survey just to monitor program operations. In summary, Bayes' rule points the way to a more manageable process of program monitoring, wherein the performance measures are themselves conditional probabilities.

Independence

We argued earlier that conditional distributions are important to planners because they embody the knowledge of particulars by which one customizes predictions. We have seen, for instance, that knowledge of age is important for predictions of chronic activity limitation. The *association* or *dependence* among attributes aids our predictions. Sometimes, of course, we understand a phenomenon so poorly (or so well) that the additional attributes we consider have very little impact on our predictions; the conditional distributions look very much like the unconditional distributions. When one attribute is essentially irrelevant to the prediction of another, we say that the attributes are *independent*. In formal terms, this means that the unconditional and conditional probabilities are identical:

$$\text{Prob } [A = i \,|\, B = j] = \text{Prob } [A = i]$$

and

$$\text{Prob}\,[B = j\,|\,A = i] = \text{Prob}\,[B = j].$$

Using this result, we can express the independence relationship in an equivalent way:

$$\text{Prob}\,[A = i \text{ and } B = j] = \text{Prob}\,[A = i\,|\,B = j] \times \text{Prob}\,[B = j]$$
$$= \text{Prob}\,[A = i] \times \text{Prob}\,[B = j].$$

This last result says that, when two attributes are independent, we can compute the probability of their joint occurrence in a very simple way: just multiply the probabilities of the individual attributes. For instance, when flipping two coins, we expect that the result for one coin will be independent of the result for the other. Hence the probability of having both coins turn up heads would be calculated as

Prob [1st coin is heads and 2nd coin is heads]
$$= \text{Prob [2nd is heads}\,|\,\text{1st is heads]} \times \text{Prob [1st is heads]}$$
$$= \text{Prob [2nd is heads]} \times \text{Prob [1st is heads]}$$
$$= 1/2 \times 1/2 = 1/4.$$

Depending on the problem at hand, independence of attributes may be either a boon or a problem. In analysis, independence always helps simplify the calculations, which is good. In prediction, independence means that attributes recorded at some cost of time and money are useless for improving predictions, which is unfortunate. Sometimes independence represents a normative condition for which we must test: the probability that a bank will grant a mortgage should be independent of race, all else equal.

How can you tell if two attributes in a table are independent? If they really are independent, you should be able to predict the entries in the table using only the marginal distributions. Returning to the example about age and activity limitation, if these two attributes were independent, then Prob [age 0–16 *and* severe limitation] should equal Prob [age 0–16] × Prob [severe limitation]. From table 2.1 the probability that a man will have both attributes is only $81/100{,}030 = 0.0008$, whereas the probability expected under the hypothesis of independence is about 20 times greater: $(32{,}080/100{,}030) \times (4{,}109/100{,}030) = 0.0164$. Thus there are many fewer young men with severe activity limitations than would be expected if age and activity limitation were independent. Roughly speaking, if you can predict the table entries fairly accurately using just the row and column marginals, then the row and column attributes can be considered independent. We will return to this topic in chapter 8.

Continuous Random Variables

To this point we have considered only discrete random variables, such as the category of activity limitation or continuous random variables lumped into discrete categories, such as the age brackets in table 2.1. However, the concepts of conditioning, Bayes' rule, and independence also apply to continuous random variables and to mixtures of discrete and continuous random variables. For instance, the distributions of incomes among rural and urban residents are different, and the conditional distributions could be determined using Bayes' rule. Suppose income is denoted by x and the probability distribution of this random variable across the entire population is denoted by $f(x)$. Then the conditional distribution of income among urban residents, which we will call $g(x \mid$ urban), will be, by Bayes' rule,

$$g(x \mid \text{urban}) \propto \text{Prob } [\text{urban} \mid x] f(x).$$

Note that only if the urban / rural mix stayed the same across all values of incomes—if Prob [urban $\mid x$] = Prob [urban]—would the conditional and unconditional distributions be identical.

We will make extensive use of this form of Bayes' rule when we address the problems of estimating a probability (chapter 6) and estimating a mean (chapter 7). In the context of estimation, we will relabel the pieces but still keep the form of the rule:

Posterior [value $\mid$ data] $\propto$ Likelihood [data $\mid$ value] $\times$ Prior [value].

Here the conditional (posterior) distribution is that of an unknown quantity (like a probability or mean) after data have been gathered, the unconditional (prior) distribution summarizes our knowledge about the value of the unknown quantity before the data have been gathered, and the two are linked by the probability (likelihood) of observing the data conditional on the value of the unknown quantity. We will return often to this statement of Bayes' rule, since it provides a graceful and consistent way to blend new data with old knowledge into a new understanding.

Summary

Conditional distributions allow us to alter our predictions about the values taken on by random variables by making use of information about other attributes of the cases. The crosstabulation is a common format for displaying the conditional distributions of discrete random variables. Bayes' rule provides a way of reversing the conditioning and conditioned attributes that can be of great help in practice, since some conditional probabilities are more readily estimated than others.

References and Readings

Schmitt. "More about Probability," chapter 2, and "Bayes' Theorem," section 3.1, pp. 34–60, 62–70.

Warner, A. "Notes on the Statistical Determination of the Causes of Poverty," *Publications of the American Statistical Association* 1 (1889): 183–201.

Warner, A. "The Causes of Poverty Further Considered," *Publications of the American Statistical Association* 4 (1894): 49–68.

Problems

2.1

Assemble the following data on Massachusetts communities into a contingency table showing the distribution of real estate assessment ratios by size of community. Communities are already aggregated into population groups (large = 50,000+, medium = 10–50,000, small = 0–10,000); choose your own groupings for assessment ratio. What is your preliminary conclusion about the nature of the relationship between the variables?

Community	Size	Assessment ratio (percent)	Community	Size	Assessment ratio (percent)
Belmont	medium	61	Monterey	small	100
Boston	large	49	Newton	large	32
Cambridge	large	38	Orleans	small	65
Carlisle	small	11	Plymouth	medium	36
Chelmsford	medium	72	Rockland	medium	55
Chicopee	large	27	Sharon	medium	78
Colrain	small	38	Springfield	large	68
Dover	small	49	Truro	small	94
Easthampton	medium	65	Waltham	large	72
Framingham	large	62	Wellesley	medium	56
Goshen	small	80	Westfield	medium	49
Greenfield	medium	74	Whately	small	13
Haverhill	medium	25	Wilbraham	medium	63
Holyoke	medium	49	Winchester	medium	60
Lawrence	large	30	Windsor	small	47
Longmeadow	medium	78	Winthrop	medium	59
Marion	small	44	Worcester	large	37
Medford	large	25	Worthington	small	30
Milton	medium	22	Yarmouth	medium	102

2.2

Refer to table 2.1. What is the probability that a working-age male will have a severe activity limitation? What is the probability that a boy 16 or younger will have an activity limitation greater than a man 65 or older (Hint: condition on the boy's degree of limitation, then combine)?

2.3

Registrants at the five neighborhood health centers in Cambridge, Mass., reported their mode of access to the center as shown in the table that follows. Make a new table showing the difference in each cell between the actual count and the count expected if mode of access were independent of facility (round to the nearest whole number). If you had to guess which center was most accessible by public transit, which center would you choose?

	Health center				
Mode	Donnelly	Windsor Street	Riverside	Cambridgeport	Fitzgerald
Private auto	62	37	40	56	86
Walking	124	219	87	111	93
Public transit	35	19	9	11	27
Other	11	8	3	5	4

2.4

Many planning decisions are influenced by population surveys. Unfortunately, surveys can be very misleading if the problem of *nonresponse bias* is severe. Suppose the survey has a yes/no answer. Since each individual will have either a yes or no position and may or may not respond to the survey, the population can be divided into four groups (assume for simplicity that no one is undecided and no one lies):

		Position on issue	
		Yes	No
Participation	Response		
in survey	Nonresponse		

a. Use Bayes' rule to show the relationship between the true probability that an individual holds the "yes" view and the probability that an individual responds, the probability that a respondent says "yes," and the probability that someone with a "yes" view is a respondent.

b. If the respondents favor "yes" by a two to one majority but only one-third of the potential respondents participated in the survey, graph and comment on the relationship between the true probability of a "yes" view and the probability that someone with a "yes" view responds.

c. If those who hold "yes" views are four times as likely to respond as those who hold "no," what is the true probability of a "yes" view?

3 Descriptive Statistics: Portable Summaries of Probability Distributions

The field of statistics is commonly divided into two parts: descriptive and inferential statistics. *Descriptive statistics* addresses itself to summarizing in brief form the information contained in a probability distribution. *Inferential statistics* addresses the leap from what one sees in a sample of cases to what may be true of the full set of cases. Our approach to descriptive statistics will focus on two questions arising from our general concern for prediction:

1. What is a typical value that we can expect a random variable to take on?
2. How much variation can we expect away from this typical value?

Of course, if we knew the full probability distribution of a random variable, we could simply read off the probability (for discrete random variables) or the relative likelihood (for continuous random variables) of any possible value. Sometimes though, we do not wish to carry around with us the full probability distribution and prefer portable summaries for convenience. At other times we may only have a few scraps of information to work with and cannot hope to construct a histogram that accurately describes the probability distribution, so we must settle for brief summaries until more data become available. Finally, it happens that some probability distributions can be described by neat mathematical formulas containing parameters whose values are either the answers to the two questions or can be derived from those answers (see the discussion of the Gaussian distribution in chapter 4). Thus for any of three reasons—desire for portability, lack of extensive data, or formulaic description of a probability distribution—we may forsake histograms in favor of more compact descriptions of random variables.

Levels of Measurement

The way we answer our two questions (What value is expected? How far will the actual value lie from the anticipated value?) will depend on the sense in which it can be said that one value is far from another. If our data are personal incomes, the differences between people have a clear numerical meaning, but if our data are rankings by degree of loyalty to the mayor, we can only put the individuals in order, not determine their spacing, and if the data are categories, such as the individuals' religious preferences, we cannot even order the list in a natural way. These differences in the ability to distinguish the distance between two values of a random variable (income, loyalty, or religion) are described as differences in *level of measurement*. Statistical techniques are designed for use with data of a particular level of measurement in much the same way that cars are designed for particular types of gasoline. It will do in a pinch to use the wrong type of gas in your car, but chronic mismatching can lead to trouble.

Consider the following three discrete random variables:

• Form of government: open town meeting, representative town meeting, city council.

• Citizen satisfaction: very dissatisfied, dissatisfied, satisfied, very satisfied.

• Pupil-teacher ratios: 10/1, 11/1, 12/1, 13/1, 14/1, . . .

You can sense the differences. The first random variable, form of government, is *nominal*: the values taken on are unordered categories (sometimes this type of variable is called "categorical"). Sex, race, religion, city of origin are all commonly treated as nominal variables. The second variable, citizen satisfaction, is *ordinal*: there is a natural order to the categories but no inherent numerical scale to the possible values. We cannot necessarily say that someone very dissatisfied is twice (or three times) as dissatisfied as someone who is merely dissatisfied. The third variable, pupil-teacher ratio, is *metric*: we can legitimately do arithmetic with it. It is very common for planners to confront nominal or ordinal data.

Is a variable born to just one level of measurement, doomed to live out its life as nominal, ordinal, or metric? Not necessarily. You can always downgrade a variable if you feel you must, although this inevitably means some information is lost in the process. For instance, data on distances of journeys to work may be (1) expressed as metric data in kilometers, (2) grouped into the ordered classes of very long, long, medium, and short trips, or (3) converted into a nominal list of name of city of origin. On the other hand, you can also upgrade a variable, but only by imposing your own (perhaps arbitrary) order or metric on the data. For instance, suppose you are studying a community's sense of citizen involvement and wish to link citizen attitudes with form of government. Ordinarily, "form of government" would be treated as a nominal variable, but you may decide that there is a natural ordering to the forms of government, with open town meeting most participatory and city council least participatory; if so, you could treat "form of government" as an ordinal rather than nominal variable. You might even think that there is a metric variable "level of participation" that takes the value 10 for open town meeting, 6 for representative town meeting, and 2 for city council.

Planners in an introductory statistics course are often unaware of the turbulence in the field they are studying. In statistics there are two schools of thought about levels of measurement. Those who are bold (rash?) advocate making "wisely arbitrary" judgments to upgrade data to the metric level so that the most powerful techniques may be applied. Those who are cautious (prudent?) hold that arbitrary scales are unpersuasive and that there should be a strict match between level of measurement and technique. Once again we recognize the importance of stylistic differences in statistical work. The lesson for the planner confronting data is that judgment counts as much here as elsewhere.

We continue by offering three answers to our two general questions, indicating in each case the level(s) of measurement with which the descriptive statistics are associated.

Measures of Central Tendency

Measures of central tendency represent typical values of random variables. The measures available for use depend on the level of measurement of the data, as shown in table 3.1.

Table 3.1
Appropriateness of some measures of central tendency for each level of measurement

Measure of central tendency	Level of measurement		
	Nominal	Ordinal	Metric
Mode	OK	OK	OK
Median	No	OK	OK
Mean	No	No	OK

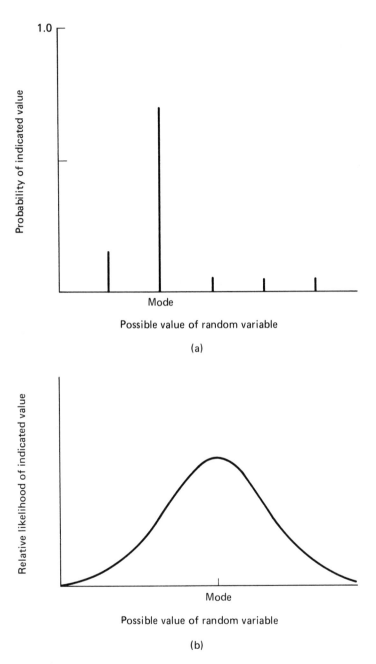

Figure 3.1
The mode for discrete and continuous probability distributions

The *mode* is that value of the random variable which is most likely to occur. Graphical indications of the mode for discrete and continuous random variables are shown in figure 3.1. The mode is useful at all three levels of measurement and is the only choice available for nominal data. One warning: the mode may not be unique for a particular distribution; in other words, two or more values of a discrete random variable may share the largest probability (see figure 1.3), or the highest part of a continuous probability distribution may be flat.

The *median* of a continuous random variable is that value which divides the probability distribution into two parts of equal area as shown in figure 3.2. The median of an odd-numbered set of metric data is the middle value:

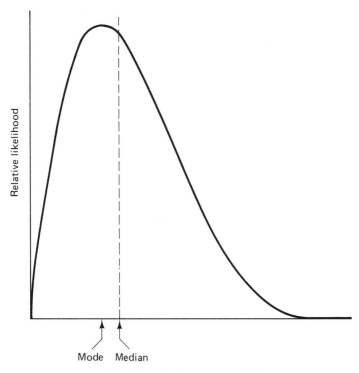

Figure 3.2
The median and mode of a continuous probability distribution

1.2   3.8   5.2   6.1   7.5
Median = 5.2.

The median of an even-numbered set of data is, by convention, the average of the two middle values:

1.2   3.8   5.2   6.1   7.5   7.6

$$\text{Median} = \frac{5.2 + 6.1}{2} = 5.65.$$

The median of an ordinal random variable is that value M for which both Prob [random variable $\geq M$] ≥ 0.5 and Prob [random variable $\leq M$] ≥ 0.5. For instance, in the following distribution of rankings the median value is "good," since Prob [ranking $\leq$ good] $= 0.85$ and Prob [ranking $\geq$ good] $= 0.55$.

Value	Probability
Excellent	0.15
Good	0.40
Fair	0.35
Poor	0.10
Sum =	1.00

The median requires at least an ordering of the possible values of the random variable, so it cannot be used with nominal data. For any given set of data, the median is less sensitive than the mean to extremely low or high values (for example, the median of 1.7, 2.1, and 3.4 is 2.1, as is the median of 1.7, 2.1, and 9,999), so it is often used to protect against undue influence by extreme values; for this reason the U. S. Census reports median housing values. Of course, for some purposes you may wish to be sensitive to extreme values; in such cases you would prefer to use the mean rather than the median.

The *mean* is the most commonly used summary measure, although it can properly be used only with metric data. The mean, unlike the mode and the median, can be easily computed from a formula. If you begin with a list of the values of a random variable X taken on by each of the N cases of interest, then the mean of X, usually denoted μ (Greek letter mu), is simply the arithmetic average

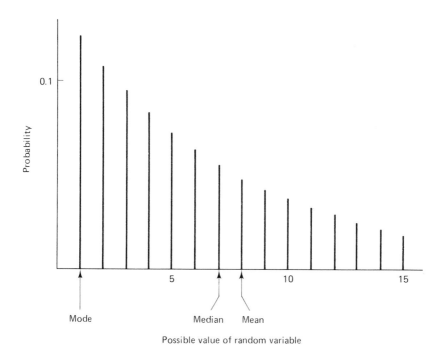

Figure 3.3
The mean, median, and mode of a discrete probability distribution

$$\mu = \sum_{i=1}^{N} X_i / N.$$

If you begin with a probability distribution for X having I possible values, then

$$\mu = \sum_{i=1}^{I} X_i \, \text{Prob} \, [X = X_i].$$

Figure 3.3 indicates the mean of a particular discrete distribution and compares it to the mode and median.

Incidently, since the mean, median, and mode can all be computed for any continuous distribution (assuming there is a unique mode), and since a distribution can be conditional, it follows that there are conditional means, conditional medians, and conditional modes. The technique of regression analysis (see chapters 9 and 10) is the most common method of using attribute data to make conditional predictions of central tendency.

Three problems seem to haunt planners when introduced to mean / mode / median. The first is that they become confused by the unfortunate fact that all three words begin with the letter "m." The second is that they cannot determine unambiguous rules for selecting one or the other measure. These two problems would doubtless seem even worse if the planners were commonly told of the fourth and fifth "m's": mid-range (the average of the smallest and largest values) and mid-interquartile range, which also provide estimates of central tendency. But perhaps the main difficulty caused by the first two problems is that they give rise to the third, which is that the planners lose sight of why measures of central tendency are valuable. The most common reason is simple: it is usually easier to carry around a measure of central tendency (or two or three) than to carry around an entire probability distribution. If you have room enough to carry around the full distribution, by all means do, for the distribution embodies more information about the random variable, but a measure of central tendency can carry a good deal of information very concisely.

Measures of Dispersion

Once we know, by dint of a mean, median, or mode, where most of the action occurs in a probability distribution, we should next inquire how much in error we will be if we use the measure of central tendency as a prediction of the value that the random variable will take on. Only rarely (if ever) will we be exactly right in such a prediction; what matters is whether we will be close enough for our purposes. If we cannot carry around the entire probability distribution, we can summarize our uncertainty about the prediction in terms of a *measure of dispersion*. We will consider here three measures of dispersion: the entropy, the range, and the standard deviation, which are matched to the levels of measurement in table 3.2.

The *entropy* is an index of uncertainty, representing in a quantitative way how well we can predict which value a random variable will take on. In planning problems, entropy has value as an index of mix, variety, or

Table 3.2
Appropriateness of some measures of dispersion for each level
of measurement

Measure of dispersion	Level of measurement		
	Nominal	Ordinal	Metric
Entropy	OK	OK	OK
Range	No	OK	OK
Standard deviation	No	No	OK

heterogeneity: mix of races, variety of land uses, heterogeneity of tenants. While entropy is defined for continuous metric variables, we will consider only discrete random variables in this discussion. Suppose we study abutters' responses to a proposed zoning variance; the possible categories of response are for, against, and no opinion. In a survey we might find the percentage of respondents in each category. When will we know most about what stand a respondent will take? Simply when the results are unanimous. When will we know least? When all of the three possible responses are equally likely. When there is unanimous agreement, we have no uncertainty in predicting the view of any given respondent, and the entropy of the probability distribution of responses will be zero. On the other hand, when all possible responses are equilikely, we have maximal uncertainty about the opinion of any given respondent, and the entropy of the distribution reaches its peak to reflect this. Thus the more concentrated the distribution, the lesser its entropy; the more diffuse the distribution, the greater its entropy. For instance, note how the entropy increases as the distribution of responses gets flatter in figure 3.4. In chapter 8 we will use the reduction in entropy of a distribution arising from using additional information about each case, as a measure of the "power" of the additional information.

The definition of the entropy of a discrete probability distribution with I possible values X_i is

$$E = -\sum_{i=1}^{I} \text{Prob} [\text{value} = X_i] \times \log \text{Prob} [\text{value} = X_i].$$

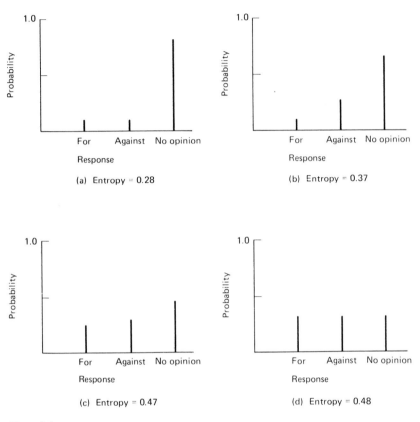

Figure 3.4
The entropy of several discrete probability distributions

(Note that, when Prob [value $= X_i$] $= 0$, we define Prob [value $= X_i$] $\times$ log Prob [value $= X_i$] $= 0$.) For example, the entropy of the third distribution in figure 3.4 is computed as

$$E = -[0.25 \log 0.25 + 0.30 \log 0.30 + 0.45 \log 0.45]$$
$$= -[(-0.15) + (-0.16) + (-0.16)] = 0.47.$$

The entropy of a distribution with I possible values ranges from a minimum of 0 to a maximum of log I. Since the log of any probability is a negative quantity, the formula contains a negative sign to make entropy a positive quantity and accomodate the average person's distaste for negative numbers.

The *range* for metric data is simply the difference between the largest and smallest occuring values, and for ordinal data is the number of categories spanned (for example, from "very satisfied" to "no opinion"). There are a cluster of statistics and statistical procedures based on the range, but it is most often used as a quick substitute for, or supplement to, the standard deviation as a measure of the width of a distribution.

The *standard deviation* is the most commonly used of the three measures of dispersion, although it requires metric data. It has a useful interpretation in terms of prediction errors. We noted earlier that the mean might be used as a kind of simple-minded prediction of the value that a random variable will take. If the mean is so used, the standard (in the sense of "typical") deviation (in the sense of "error") is a measure of the typical size of the error produced by this method of prediction.

If you begin with a list of the values of a random variable X taken on by each of the N cases of interest, then the standard deviation of X, usually denoted σ (Greek letter sigma), is

$$\sigma = \sqrt{\frac{\sum_{i=1}^{N} (X_i - \mu)^2}{N}}$$

where μ is the mean of X. If you begin with a discrete probability distribution of X having I possible values, then

$$\sigma = \sqrt{\sum_{i=1}^{I} (X_i - \mu)^2 \, \text{Prob [value} = X_i]}.$$

For instance, if a check of sales prices for all homes sold in a community during one year reveals a mean sales price of $\mu = \$43,000$ with a standard deviation of $\sigma = \$12,000$, then a reasonable estimate of the sales price of any one of those homes is $43,000, and such an estimate will be in error by about $12,000 on average.

There are two commonly used quantities that derive from the standard deviation. One is the *variance*, which is just the square of the standard deviation. The other is the sample standard deviation, which has a slightly different formula and is used when we have information not on all N cases of interest but only on a sample of size $n < N$. We will use the sample standard deviation in chapter 7.

Coefficient of Variation

Our concern with probability distributions may have impressed you with the extra effort required to deal with randomness. You should be aware, then, that you may be able to ignore the randomness and treat random variables as though they were constants when the measure of dispersion is much less than the measure of central tendency. How much less depends, as always, on the problem at hand, but if the mean is $100 and the standard deviation only 10¢, then the randomness in the problem may be insignificant. In this way one can combine measures of central tendency with measures of dispersion to assess the degree of randomness. A formal way to do this with metric data is to compute the *coefficient of variation*, defined as the ratio of the standard deviation to the mean:

$$CV = \frac{\sigma}{\mu}.$$

As a device for summarizing the degree of randomness, the coefficient of variation has one advantage over the standard deviation: it takes account of the scale of the problem. Thus a standard deviation of $1,500 in personal income in a community may seem large, but if the mean personal income in the community were $45,000, then the relative variability expressed in the coefficient of variation ($CV = $1,500/$45,000 = 0.03$) would be very small. In such a case it is rather safe to use the community's mean income as a prediction of income in an individual case, since the prediction would be in error by only about 3 percent on average. On the other hand, if the community's mean income were only $6,000, then the coefficient of variation would equal 0.25, and using the community's mean income as a prediction for an individual case would be in error by about 25 percent on average. The coefficient of variation provides a simple way of summarizing the salience of the variability in a quantity.

Standardized Random Variables

Suppose you are told that the unemployment rate in a city is 6.7 percent. How do you react to such a fact? You might have some reaction based on the absolute magnitude; for instance, you may be displeased by any unemployment rate above 5 percent. On the other hand, you might also want to react in a relative way: is 6.7 percent high or low relative to comparable cities? If it is high, is it slightly high or very high? *Standardized random variables* are shifted and scaled versions of the original random variables that permit you to answer such relative questions at a glance.

Formally, the standardized random variable Z corresponding to the (original) unstandardized random variable X is defined as

$$Z = \frac{X - \mu}{\sigma},$$

where μ and σ are the mean and standard deviation of X, respectively. The random variable X is first shifted by subtracting its mean, then scaled by dividing by the standard deviation. Subtracting the mean gives us a quick way of telling whether any given value of X is high or low, since values of X greater than the mean correspond to positive values of Z, while values of X less than the mean correspond to negative values of Z. Dividing by the standard deviation gives us a way of telling how extreme the deviation of X is from the mean, since a typical deviation is one standard deviation. Thus a value of $Z = -4.0$ corresponds to a value of X not only below the mean value but very far from the mean relative to the dispersion of all the cases. A value of $Z = +0.62$ corresponds to a value of X above the mean value but not very far from it compared to the overall scatter of the cases. A value of $Z = +2.2$ corresponds to a rather high value of X. Converting the original data into standardized form permits rapid interpretation of the relative position of each case. Returning to the preceding example, suppose the unemployment rate in the city of interest and all comparable cities has a mean value of 8.2 percent and a standard deviation of 0.5 percent. In this case the standardized unemployment rate in the city of interest is

$$Z = \frac{6.7 - 8.2}{0.5} = -3.0.$$

Thus, although that city may have an unemployment rate judged to be too high in normative terms, it is doing very well relative to comparable cities.

The judgment about whether a standardized random variable $Z = +2.0$ is "rather high," "high," or "very high" depends on the specific probability distribution of the unstandardized random variable X. If we know nothing at all about X except that it is positive, a general result from probability theory known as *Chebyshev's inequality* conservatively assures us that the probability that Z is outside the range -2.0 to $+2.0$ is at most 0.25, and the probability that Z is outside the range -3.0 to $+3.0$ is at most 0.11. In this sense a standardized value of $+2.0$ is "high" and a value of $+3.0$ is "very high." If we know in addition that the random variable X has a distribution symmetric about its mean, then Chebyshev's inequality assures us that standardized values more than 2 or 3 standard deviations from the mean occur with probabilities less than 0.11 and 0.05, respectively. (If we know further that the random variable X has a Gaussian distribution, see chapter 4, then we know that the probabilities that Z is more than 2 or 3 standard deviations from the mean are exactly 0.046 and 0.003, respectively, so the Chebyshev bounds may be rather conservative.) Standardized random variables are useful in their own right and in calculating probabilities for Gaussian random variables.

Summary

It is often useful to summarize briefly the main features of a probability distribution. The typical value taken on by a random variable is summarized by a measure of central tendency, such as the mode, median, or mean. The typical variation away from the typical value is summarized by a measure of dispersion, such as the entropy, range, or standard deviation. The choice of summary measures will depend on the level of measurement of the data: nominal, ordinal, or metric. The degree of randomness can be summarized for metric data using the coefficient of variation, which can be interpreted as the proportional error arising from using the mean value of a random variable to represent the value of an individual case. Standardizing a random variable facilitates interpretation of the relative position of one case compared to all cases.

References and Readings

Cesario, F. S. "A Primer on Entropy Modelling." *American Institute of Planners Journal* 41 (1975): 40–47.

Hodge, G. "Use and Mis-use of Measurement Scales in City Planning." *American Institute of Planners Journal*, 41 (1963): 112–121.

Mosteller and Tukey. "Robust and Resistant Measures," chapter 10, pp. 203–219.

Tomlin, J., and S. Tomlin. "Traffic Distributions and Entropy." *Nature* 220 (1968): 974–976.

Wilson, A. *Entropy in Urban and Regional Modeling*. London: Pion Limited, 1970.

Problems

3.1

McClure gathered data on mean values of nine attributes within the neighborhoods of Cambridge, Mass. He was not able to obtain information on four of the attributes of neighborhood 12. Use the information in each column having a missing value to fill in a reasonable estimate of that value. How large an error might you expect to make in each case, using this procedure?

Analysis of neighborhood characteristics

Neighborhood	Household income	Rent per apartment	Rent per room	Rooms per apartment	Persons per room
1	11,481	152.98	35.35	4.49	0.681
3	9,392	197.42	57.99	4.11	0.679
4	8,074	154.70	36.64	4.29	0.640
5	8,353	231.64	66.02	3.54	0.723
6	11,658	220.74	61.30	4.06	0.539
7	9,295	202.47	46.34	4.68	0.504
8	10,798	248.69	56.89	4.57	0.490
9	7,730	231.37	63.07	4.03	0.538
10	10,989	256.56	51.04	5.09	0.439
11	8,135	190.93	46.55	4.57	0.635
12	—	195.00	33.50	6.00	0.542
13	8,622	168.26	33.33	5.00	0.662

Neighborhood	Length of residence	Percentage nonwhite	Percentage student	Percentage 65+
1	3.68	18.5	22.3	5.2
3	3.34	10.7	21.7	9.0
4	2.64	25.2	34.2	5.5
5	3.40	10.0	33.8	21.7
6	2.54	17.2	43.2	4.2
7	2.91	27.3	45.6	4.3
8	3.09	15.3	47.6	9.6
9	2.97	26.5	42.2	9.2
10	3.06	26.7	34.4	8.9
11	2.41	43.3	42.3	4.1
12	4.30	—	—	—
13	2.71	28.0	30.6	7.6

3.2

Do Cambridge neighborhoods show greater relative diversity in percentage nonwhite or in percentage student?

3.3

Neighborhood 6 has a relatively high mean rent per room and a relatively low mean length of residence. Which of these two attributes is more extreme compared to the values taken on in other neighborhoods?

3.4

Presented below are data on the distribution of taxable property valuation across property categories in three Massachusetts communities. Which community has the property tax base of greatest diversity?

Percentage of total taxable property valuation (1975)

Community	1–3 family homes	4+ family homes and mixed residential / commercial	Commercial	Industrial	Personal	Other	Total
Belmont	84	5	8	0.5	2	0.5	100
Cambridge	25	16	24	16	17	2	100
Boston	22	15	37	10	14	2	100

3.5

Hadaway studied the return on investment in agricultural land in ten regions across the United States. He reported the mean and standard deviation of the annual percentage price change less real estate taxes over the period 1955 to 1974, as shown in the table. Comment on this conclusion drawn by Hadaway:

Simply plotting the data ... it becomes obvious that some individual regions have represented better risk-return combinations than others. Regions such as the Corn Belt and Northern Plains are clearly dominated by either the Lake States of the Southeast.

Return and variability data for portfolio possibilities (1955 to 1974)

Region	Mean return[a] (percent)	Standard deviation (percent)
1. Northeast	6.438	4.312
2. Lake states	5.019	4.837
3. Corn belt	5.226	5.891
4. Northern plains	5.037	5.805
5. Appalachia	7.194	3.860
6. Southeast	9.121	5.263
7. Delta states	7.814	4.195
8. Southern plains	6.805	4.796
9. Mountain	6.183	4.970
10. Pacific	3.656	3.225
All regions (U.S.)	6.856	4.506

[a] Simple mean of annual percentage price change less real estate taxes.

4 Some Probability Distributions of Special Interest to Planners

Many of the probability distributions you will meet in practice will be empirical, arising from data obtained in studies of particular urban systems. It will nevertheless be very useful to know several theoretical probability distributions because the theoretical distributions may closely approximate the empirical distributions and are often easier to work with and because certain theoretical distributions help us understand the process of taking samples from populations. There are perhaps two dozen common theoretical distributions. We will pay special attention to five of the most useful: the *hypergeometric*, the *binomial*, the *beta*, the *Poisson*, and the *Gaussian*.

The Hypergeometric Distribution

This distribution is very useful for addressing one of the most basic prediction problems faced by planners: estimating a proportion by sampling from a finite population. Suppose your task is to determine how many of a certain category of welfare recipients are receiving fewer benefits than they are entitled to receive. A local welfare office may have 100 such cases on its rolls. Since it is very time consuming to check even one case for proper benefit level, you despair of checking all 100 before your report is due. To get some sense of the error rate without spending forever in the checking, you might select 20 files at random and check only those. From a *population* of 100 cases you select a *sample* of 20. The chances are good that if there are no underpayments in your sample of 20, there will not be very many cases of underpayment in the full set of 100. In chapter 5 we will return to this problem to obtain the probability distribution of the number of underpaid cases in the population, conditional on the number of underpaid cases found in the sample. In this chapter we will address the obverse task: estimating the probability distribution of the number of underpaid cases in the sample, conditional on the number of underpaid cases in the population. We begin with a small example.

Consider a population of 5 cases: 3 A's and 2 B's (the B's might represent the underpaid cases in the welfare example). Suppose we randomly select 2 of the 5 cases. What is the chance that the sample of 2 would contain one B and one A? Shown in table 4.1 are all 10 possible samples of 2 cases from a population of 5 cases. In a *simple random sample* each

Table 4.1
The 10 possible samples of size 2 from a population having 2 cases of one type and 3 cases of another

Case number:	1	2	3	4	5	
Case type:	*B*	*B*	*A*	*A*	*A*	Number of *B*'s in sample
Sample: 1	*X*	*X*				2
2	*X*		*X*			1
3	*X*			*X*		1
4	*X*				*X*	1
5		*X*	*X*			1
6		*X*		*X*		1
7		*X*			*X*	1
8			*X*	*X*		0
9			*X*		*X*	0
10				*X*	*X*	0

of these 10 samples has an equal chance of appearing. Six of the 10 equilikely samples contain one *A* and one *B*, so Prob [one *A* and one *B* in sample | $3A$'s and $2B$'s in population] = 0.60. We can summarize the results in table 4.1 in the following distribution of the number of *B*'s in the sample

Prob [0*B*'s in sample of 2 | 2*B*'s in population of 5] = 0.30
Prob [1*B* in sample of 2 | 2*B*'s in population of 5] = 0.60
Prob [2*B*'s in sample of 2 | 2*B*'s in population of 5] = 0.10

$$\text{Sum} = 1.00.$$

We might use the same approach to analyze the problem of the welfare office with its population of 100 cases and sample of 20, but the list of all possible samples would be so huge that the effort would be impractical. Luckily, there is a mathematical formula that can express the hypergeometric probabilities concisely. In general, we have a population of N cases, of which M have some attribute of interest and $N - M$ do not. We draw a simple random sample of n cases, of which m have the attribute and $n - m$ do not. The random variable is m, the number of cases in the sample having the attribute of interest. Its probability distribution, conditional on the size of the sample (n), the size of the population (N), and the number of cases with the attribute in the population (M), is

$$\text{Prob}\,[m\,|\,n,\,N,\,M] = \frac{\dbinom{M}{m}\dbinom{N-M}{n-m}}{\dbinom{N}{n}},$$

$0 \le m \le$ minimum of M and n.

The expression $\binom{X}{Y}$ is known as a *binomial coefficient* and represents the number of different ways of forming sub-groups of size Y out of a group of size X

$$\binom{X}{Y} = \frac{X!}{(X-Y)!\,Y!}$$

$$= \frac{(X)(X-1)(X-2)\ldots(3)(2)(1)}{(X-Y)(X-Y-1)(X-Y-2)\ldots(3)(2)(1)(Y)(Y-1)(Y-2)\ldots(3)(2)(1)}.$$

For example, the number of different ways of forming samples of size $n = 2$ from a population of size $N = 5$ is

$$\binom{5}{2} = \frac{5!}{(5-2)!\,2!} = \frac{5!}{3!\,2!} = \frac{(5)(4)(3)(2)(1)}{(3)(2)(1)(2)(1)} = 10.$$

The symbol $x!$ is read "x factorial."

Now consider the form of the hypergeometric distribution. The denominator term $\binom{N}{n}$ represents the number of different ways to form samples of size n from a population of size N. Since each possible sample is equilikely, and there are $\binom{N}{n}$ of them, the probability of obtaining any particular sample is $\frac{1}{\binom{N}{n}}$. But there may be many samples that have the same result, a total of m cases with the attribute and $n - m$ without. For instance, the small example in table 4.1 shows six samples all having one A and one B. Therefore the numerator of the hypergeometric formula must be the number of different ways of obtaining the given division of the sample. Indeed, there are $\binom{M}{m}$ different ways to select m cases from among the M with the attribute and $\binom{N-M}{n-m}$ different ways to select $n - m$ cases from among the $N - M$ without the attribute. Since any combination of m and $n - m$ will do, there must be $\binom{M}{m}\binom{N-M}{n-m}$ different ways to constitute a sample divided into m cases with and $n - m$ cases without the attribute. Turning again to table 4.1, there are $\binom{M}{m} = \binom{3}{1} = 3$ ways to get one A in the sample: either case 3, case 4, or case 5 will do. Likewise there are $\binom{N-M}{n-m} = \binom{2}{1} = 2$ ways to get one B in the sample: either case 1 or case 2 will do. Thus there are $\binom{M}{m}\binom{N-M}{n-m} = \binom{3}{1}\binom{2}{1} = 3 \times 2 = 6$ different ways to get one A and one B, and these represent 6 of the $\binom{N}{n} = \binom{5}{2} = 10$ possible samples. Hence the probability of obtaining exactly one B in a sample of size $n = 2$ is $6/10 = 0.60$.

To return to the example of the welfare office, suppose there are $N = 100$ cases, of which $M = 10$ are underpaid. Then the probability of finding $m = 0$ underpaid in a sample of size $n = 20$ is

$$\text{Prob}\,[m = 0 \,|\, n = 20,\, N = 100,\, M = 10]$$

$$= \frac{\binom{10}{0}\binom{100-10}{20}}{\binom{100}{20}} = \frac{\binom{10}{0}\binom{90}{20}}{\binom{100}{20}}$$

$$= \frac{\left(\dfrac{10!}{10!0!}\right)\left(\dfrac{90!}{70!20!}\right)}{\left(\dfrac{100!}{80!20!}\right)} = \frac{80!90!}{70!100!}$$

$$= \frac{(80)(79)(78)\ldots(71)}{(100)(99)(98)\ldots(91)} = 0.0951.$$

(It is conventional to take $0! = 1.0$.)

The entire probability distribution of the number (and proportion) of underpaid cases in the sample is shown in table 4.2. Note that the modal count of underpaid cases in the sample of 20 is $m = 2$, corresponding to a 10 percent underpayment rate, which equals the actual underpayment rate in the population ($M/N = 10/100 = 0.10$). Thus the most likely sample result is the right answer. However, other sample results are possible, as shown in table 4.1. Whenever we sample, we run the risk of obtaining a sample result more or less different from the true result for the entire population.

The hypergeometric distribution has a mean equal to Mn/N and a standard deviation equal to

$$\sqrt{\frac{(N-n)(N-M)Mn}{N^2(N-1)}}.$$

Table 4.2
Distribution of the number of underpaid cases in a random
sample of 20 drawn from a population of 100 containing 10
underpaid cases

Number of underpaid cases in sample	Hypergeometric probability of this sample result
0	0.0951
1	0.2679
2	0.3182[a]
3	0.2092
4	0.0841
5	0.0215
6	0.0035
7	0.0004
8	0.0000+
9	0.0000+
10	0.0000+
11+	0[b]

[a] Note that the most likely sample result is 2 underpaid cases,
corresponding to a fraction underpaid of 2/20 = 0.10, which is
the true fraction underpaid in the entire population of 100
cases.
[b] Since there are only 10 underpaid cases in the entire population, there cannot be more than 10 underpaid cases in the
sample.

Note that, as the sample size n approaches the population size N, the
standard deviation shrinks toward 0. This is as it ought to be, since as we
take larger and larger samples we know more about the population and
have less chance of obtaining a sample result unrepresentative of the
entire population. The fact that the proportion m/n of sample cases having
the attribute in question has as its typical value the true proportion in
the population M/N reassures us that, on average, a simple random sample will accurately reflect the population from which it is drawn.

The Binomial Distribution

Consider a sample case with probability P of having a certain attribute
and probability $1 - P$ of not having the attribute. For instance, the case
may be an individual asked to serve on an architectural jury to decide

between two competing designs, and the attribute may be forming a favorable impression of one of the designs. For now we will assume that the probability P is known. The procedure of asking the individual jury member to approve or disapprove a design is a *Bernoulli trial*: the result will be positive with probability P or negative with probability $1 - P$. It is assumed that the individual's response will be independent of the responses of the other jurors. An independent trial with a yes/no type outcome is a Bernoulli trial. A single coin flip is the archetypical Bernoulli trial. Let us use "success" and "failure" as general designations for the two possible outcomes of a Bernoulli trial. We may specialize these to live/die, win/lose, approve/disapprove, move/stay, or any other dichotomy, as appropriate for any particular problem.

The *binomial distribution* describes the total count of cases having an attribute after a number of identical Bernoulli trials. In the jury example, the total count of jurors approving of the design would be a random variable with a binomial distribution. If there were three jurors, then the number approving the design would be a random number between 0 and 3 inclusive. The total would be the aggregation of the results of the individual Bernoulli trials. Table 4.3 lists all the possible outcomes of the three jurors' judgments, where a favorable judgment is termed a success. Since there are 3 jurors and each will make one of 2 judgments, there are $2 \times 2 \times 2 = 2^3 = 8$ possible outcomes. In general, if there were n

Table 4.3
The 8 possible outcomes of the deliberations of 3 jurors (F = failure, S = success)

Juror:	1	2	3	Number of successes	Probability of outcome
Outcome: 1	F	F	F	0	$(1 - P) \times (1 - P) \times (1 - P)$
2	F	F	S	1	$(1 - P) \times (1 - P) \times (P)$
3	F	S	F	1	$(1 - P) \times (P) \times (1 - P)$
4	F	S	S	2	$(1 - P) \times (P) \times (P)$
5	S	F	F	1	$(P) \times (1 - P) \times (1 - P)$
6	S	F	S	2	$(P) \times (1 - P) \times (P)$
7	S	S	F	2	$(P) \times (P) \times (1 - P)$
8	S	S	S	3	$(P) \times (P) \times (P)$
					Total = 1.0

Bernoulli trials there would be 2^n possible outcomes. Of course many of the outcomes are identical as far as the total count of successes is concerned; for instance, outcomes 4, 6, and 7 in table 4.3 all represent 2 successes and 1 failure. In general there are $\binom{S+F}{S}$ ways to arrive at S successes out of a total of S successes and F failures. In table 4.3 there is $\binom{3}{0} = 1$ way to arrive at 0 successes among 3 trials, there are $\binom{3}{1} = 3$ ways to arrive at 1 success and $\binom{3}{2} = 3$ ways to arrive at 2 successes, and there is $\binom{3}{3} = 1$ way to arrive at 3 successes.

Note that the eight possible outcomes in table 4.3 are not equilikely. Since the judgments are presumed to be independent, the probability of any set of three outcomes can be computed simply as the product of the three individual outcomes. Since in each trial there is a probability P of success and $1 - P$ of failure, the probability of outcome 1 (3 failures) is $(1 - P) \times (1 - P) \times (1 - P) = (1 - P)^3$, while the probability of outcome 8 (3 successes) is $P \times P \times P = P^3$. In general, the probability of any single outcome having S successes and F failures is $P^S(1 - P)^F$, and there will be $\binom{S+F}{S}$ such outcomes, so

$$\text{Prob}\,[S\,|\,P, S + F] = \binom{S+F}{S} P^S(1 - P)^F.$$

This is the formula for the binomial probability distribution. The mean of the binomial is $(S + F)P$, and the standard deviation of the binomial is $(S + F)\sqrt{P(1 - P)}$. A table of binomial probabilities is available in appendix A.

In the jury example, suppose the probability that an individual juror will approve the design is $P = 0.60$. Then

Prob [0 of 3 approve] $= \binom{3}{0}(0.6)^0(1-0.6)^3 = 0.064$
Prob [1 of 3 approve] $= \binom{3}{1}(0.6)^1(1-0.6)^2 = 0.288$
Prob [2 of 3 approve] $= \binom{3}{2}(0.6)^2(1-0.6)^1 = 0.432$
Prob [3 of 3 approve] $= \binom{3}{3}(0.6)^3(1-0.6)^0 = \underline{0.216}$
Sum $= \overline{1.000.}$

Let us contrast the binomial distribution with the hypergeometric. The hypergeometric pertains to samples drawn from finite populations having attributes fixed in advance. The binomial pertains to aggregations of Bernoulli trials wherein the attributes are unknown in advance. That the process of sampling from a finite population is not a Bernoulli trial can be demonstrated by referring to the welfare office example in which the population of 100 cases contained 10 cases of underpayment. Suppose we wanted to model the process of classifying each case "underpaid" or "not underpaid" as a Bernoulli trial. There are certainly just two possible outcomes, but the trials are by no means independent. Consider the first case drawn at random from the population of 100; there is a 10/100 probability that the first case examined will be underpaid. If it is, there is a 9/99 probability that the second case drawn will be underpaid; if it is not, then the probability is 10/99. In either event, the probability of success on the second trial is different from that on the first trial and depends on the outcome of the first trial. Thus sampling from a finite population does not constitute a sequence of independent trials.

On the other hand, if the sample is small relative to the size of the entire population, it is possible to approximate the sampling process as a sequence of Bernoulli trials. The advantage of this approximation is that tables of the binomial distribution are more readily available than tables of the hypergeometric and, in the absence of tables, the mathematical form of the binomial distribution is easier to work with. However, since it is rather common for planners to work with samples that constitute a relatively large portion of a relatively small population, it is important to be aware that the hypergeometric provides an exact way of dealing with

Table 4.4
The binomial distribution as an approximation to the hypergeometric

Number of underpaid cases in sample	Hypergeometric probability	Binomial approximation
0	0.0951	0.1216
1	0.2679	0.2702
2	0.3182	0.2852
3	0.2092	0.1901
4	0.0841	0.0898
5	0.0215	0.0319
6	0.0035	0.0089
7	0.0004	0.0020
8	0.0000+	0.0004
9	0.0000+	0.0001
10	0.0000+	0.0000+
11+	0	0.0000+

such samples, and the more commonly used binomial represents an approximation. Of course, the binomial is the proper distribution to use in problems resembling coin flips, in which the attributes are not fixed before the sample is taken.

To get a feel for the adequacy of the binomial as an approximation to the hypergeometric, consider again the example of the welfare office. If we ignore the finiteness of the population relative to the size of the sample, and if we call discovery of an underpaid case a success, then we might say that any given case sampled has a probability of success $P = 10/100 = 0.1$. Hence the probability of observing S successes in the sample of 20 is approximately $\binom{20}{S}(0.1)^S(0.9)^{20-S}$. Table 4.4 compares these approximate binomial probabilities to the exact hypergeometric. The binomial approximates best in the middle and poorly on the tails of the distribution.

The Beta Distribution

The beta distribution applies to continuous random variables restricted in range between zero and unity. Accordingly, it will play a central role in the problem of estimating a probability (chapter 6).

Let P be a random variable between 0 and 1.0. It may represent the probability that an individual will sell his house, or move to another city, or die; it may represent the probability that an individual who has received job training assistance will earn more than a similar individual without such training; it may represent the probability that a pollution source will respond positively to a new set of incentives to reduce environmental damage. In each case P represents an unknown probability about which we can ask interval questions: What is the probability that the unknown probability P exceeds 0.6? What is the probability that the unknown probability P lies between 0.14 and 0.35? One possible distribution for the unknown probability P is the *beta*, defined by the formula

$$f(P \mid S, F) = \frac{(S + F + 1)!}{S!F!} P^S (1 - P)^F.$$

The beta distribution is remarkably flexible yet simple. With just the two parameters S and F the shape of the distribution can be manipulated dramatically, as shown in figure 4.1. When $S = F = 0$, the beta distribution is *uniform*, or flat across the range 0 to 1.0. When S equals F, the beta

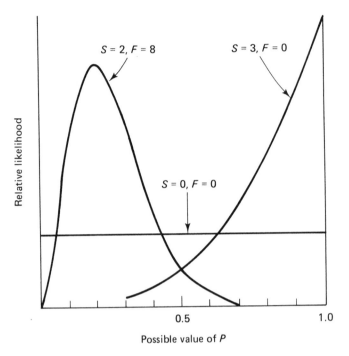

Figure 4.1
The beta distribution for various pairs of the parameters S and F

distribution is symmetrical; when S exceeds F the beta bunches to the right (in formal terms, is "positively skewed"); when F exceeds S the beta bunches to the left ("negatively skewed"). When the sum $S + F$ is small, the beta is diffuse; when $S + F$ is large, the beta is concentrated. The mode of the beta distribution occurs at $\dfrac{S}{S + F}$, the mean at $\dfrac{S + 1}{S + F + 2}$. The standard deviation of the beta is

$$\sqrt{\frac{(S + 1)(F + 1)}{(S + F + 2)^2 (S + F + 3)}}.$$

You will no doubt have noticed the similarity between the forms of the beta and the binomial distributions. In both cases, the key part of the formula is the expression $P^S(1 - P)^F$; the other terms in both cases are just normalizing constants, required to make the sum of the probabilities equal unity in the case of the discrete binomial and to make the area under the distribution equal unity in the case of the continuous beta. The real difference between the binomial and the beta concerns what is known and what is unknown. With the binomial, the probability P is known and the numbers of successes and failures in a given set of trials are unknown. With the beta, the numbers of successes and failures are known, while the probability P is unknown. Since we often care only about relative likelihoods, not absolute probabilities, we will be able to use the table of the binomial in appendix A to calculate (unnormalized) values of the beta distribution by reading it as if the parameter P were the unknown.

The Poisson Distribution

The Poisson distribution is commonly used to describe the count of events occuring at random in time or space. The number of cars passing through an intersection during a certain hour, the number of calls for emergency ambulance service during a tour of duty, and the number of fires arising in a neighborhood are all examples of random variables that have been usefully described by the Poisson distribution. These random variables are discrete and (in theory) can take on any value from zero to infinity.

Just as the binomial distribution can be thought of as the limiting case of the hypergeometric as the size of the population approaches infinity, so the Poisson can be though of as the limiting case of the binomial when the total number of Bernoulli trials approaches infinity at the same time that the chance of a success in any individual trial approaches zero. For instance, imagine a very large number of identical rural households. On any given day, there is a very small probability that any given household will migrate to a city. If we knew the total number of rural households and the probability that any one would migrate on a given day, and if each household's decision to migrate were made independently of its neighbors' decisions, then the total number migrating on a given day would have a binomial distribution. If the number of rural households were very large and the chance of any household's moving on a given day were very small, then the number of households moving on a given day would be well approximated by the Poisson distribution. Many situations of interest to planners share the three features of (1) a large number of individual cases, (2) roughly the same low probability of success in any individual trial, and (3) little or no dependence among trials: the number of auto breakdowns on an expressway at rush hour, the number of heart attack deaths per week in a county, the number of city managers fired in the U.S. during one year, the number of homes catching fire in a city during the summer, and so on. Of course it will always be true in reality that the number of cases is finite, that the cases are heterogeneous in that they do not all have exactly the same small probability of success and that outcomes in some cases will have a bearing on the outcomes of other cases; no abstract model of reality can be perfectly accurate, but the Poisson often does well as an approximation.

Unlike the hypergeometric formula, which required three parameters (n, M, and N), and unlike the binomial, which required two (P and the sum $S + F$), the Poisson formula requires only one parameter: the average number of events expected, which we will denote as A:

$$\text{Prob}\,[m\,|\,A] = A^m \exp(-A)/m!, \qquad 0 \le m \le \infty.$$

A handy way to compute successive probabilities for the Poisson distribution is to begin by computing

$$\text{Prob}\,[0\,|\,A] = \exp(-A)$$

and then use the relation

$$\text{Prob}\,[m\,|\,A] = (A/m)\,\text{Prob}\,[m - 1\,|\,A]$$

to compute the next probability from the last. For instance, if the random variable of interest is the number of calls for police service from the central business district during the noon hour and the average number is $A = 3.5$, then

$$\text{Prob}\,[0\,|\,3] = \exp(-3.5) = 0.030$$
$$\text{Prob}\,[1\,|\,3.5] = (3.5/1)\,\text{Prob}\,[0\,|\,3.5] = 0.106$$
$$\text{Prob}\,[2\,|\,3.5] = (3.5/2)\,\text{Prob}\,[1\,|\,3.5] = 0.185,$$

and so forth. A graph of the Poisson distribution with $A = 2.0$ is shown in figure 1.3. The mean of the Poisson distribution with parameter A is equal to A, and the standard deviation is equal to $\sqrt{A}$.

The Gaussian Distribution

The hypergeometric, binomial, and Poisson distributions are applied to discrete random variables. Perhaps the best known distribution for continuous random variables is the Gaussian (sometimes called the *normal* distribution or the "bell-shaped curve"). Many individual random variables have the Gaussian distribution. More importantly, sums of random variables tend to be Gaussian, which is significant since many quantities of interest to planners are aggregates: total electric power consumption in a community, total spending on tourism, total number of hospital admissions in a catchment area, and so on. A special kind of aggregate is the mean, formed by adding a number of quantities and dividing by that number. Means tend to have a Gaussian distribution even if their constituent parts do not, and one of the more common prediction problems encountered by planners is to estimate the mean value of some quantity in a population from the mean observed in a sample of that population (see chapter 7).

We require two parameters to specify a Gaussian distribution: the mean, which will center the distribution, and the standard deviation, which will determine how spread out the distribution is.

Letting μ represent the mean value of the random variable and σ the standard deviation, the formula for the Gaussian distribution is

$$f(x \mid \mu, \sigma) = \frac{1}{\sqrt{2\pi}\,\sigma} \exp\left[-\frac{1}{2}\left(\frac{x-\mu}{\sigma}\right)^2\right].$$

Figure 4.2 illustrates how the Gaussian distribution depends on the mean (which shifts it left or right) and the standard deviation (which spreads and lowers or narrows and raises the curve).

If we are willing to assert that a random variable has a Gaussian distribution, and we have estimates of its mean and standard deviation, then we can answer any of the interval questions appropriate for continuous random variables, such as, "What is the probability that the random variable will take on a value between two and three standard deviations above its mean?" In practice, we never really calculate these answers; rather, we look them up in tables of the Gaussian distribution. Those with a practical bent will wonder how large these tables must be,

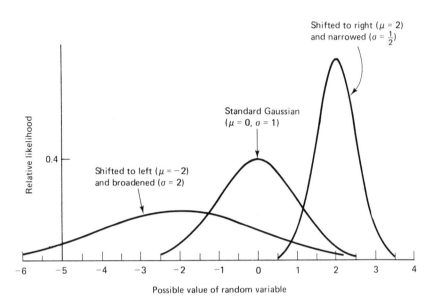

Figure 4.2
The Gaussian distribution with various values of mean μ and standard deviation σ

since we apparently need a separate table for each and every Gaussian distribution, that is, for every possible pair of mean and standard deviation, and there are an infinite number of such pairs. Actually, we can easily make do with just one master table by standardizing the random variable. Standardizing permits us to refer all questions to a single table of the standard Gaussian with mean equal to zero and standard deviation equal to unity. The procedure is as follows:

Suppose a city's yearly expenditures on snow removal have a Gaussian distribution with mean $\mu = \$60,000$ and standard deviation $\sigma = \$12,500$ (How we would estimate these numbers is an important problem in itself; assume for now that they are known). If the city manager decides to set aside a contingency fund of $80,000 for snow removal, what is the chance that the city's actual expenses will exceed the amount set aside in the contingency fund?

We can answer the question graphically easily enough: the answer is the area to the right of $80,000 under the graph of the Gaussian distribution of snow removal costs (recall that the area under the entire curve must be unity), see figure 4.3 a and b. You should always draw a curve like that in the figure to help you visualize the problem, but it does not need to be drawn very precisely because you will not be counting squares beneath the curve to find the area and thus the probability. Instead, note that the area to the right of $80,000 in the unstandardized case is exactly the same as the area to the right of the standardized version of $80,000 under the standard Gaussian distribution. Since the standardized value of a random variable is formed by subtracting the mean, then dividing by the standard deviation, the standardized value of $80,000 is

$$Z = \frac{\$80,000 - \$60,010}{\$12,500} = 1.60.$$

Hence

Prob [snow removal cost > $80,000]
 = Prob [standard Gaussian > 1.60]
 = Prob [$Z > 1.60$].

From appendix B which contains a table of the standard Gaussian distribution we obtain

Prob [$Z > 1.60$] = 1.0 − Prob [$Z \le 1.60$] = 1.0 − 0.945 = 0.055.

Thus there is about a 5 percent chance that the $80,000 snow removal contingency fund will prove inadequate.

As a second example, suppose the ratio of actual cost to estimated cost for a new shopping center is found to be normally distributed with mean 1.15 (an average cost overrun of 15 percent) and standard deviation 0.20. What is the chance that a given shopping center will actually be built for between 20 and 30 percent under its estimated cost?

First, as always, draw the picture, as in figure 4.3c. The answer is the area under the Gaussian curve of cost ratio between 0.7 and 0.8. The answer is also the area under the standard Gaussian curve between the standardized equivalents of 0.7 and 0.8,

$$Z_{0.7} = \frac{0.7 - 1.15}{0.20} = -2.25,$$

$$Z_{0.8} = \frac{0.8 - 1.15}{0.20} = -1.75.$$

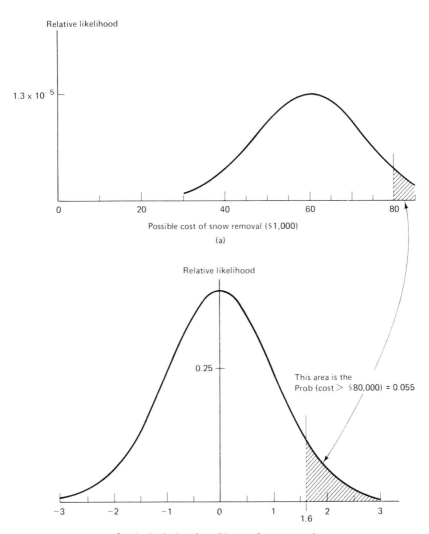

Relative likelihood

1.3 x 10⁻⁵

Possible cost of snow removal ($1,000)

(a)

Relative likelihood

0.25

This area is the
Prob (cost > $80,000) = 0.055

1.6

Standardized value of possible cost of snow removal

(b)

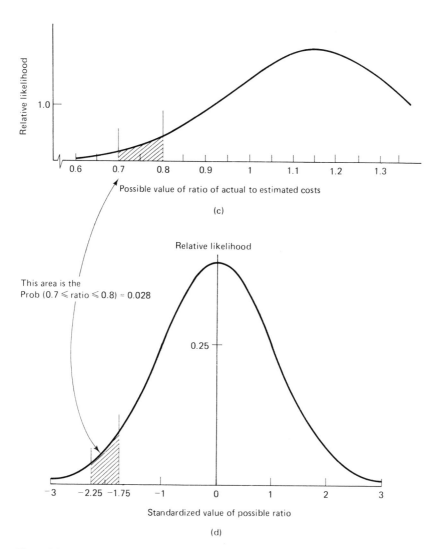

Figure 4.3
Uses of the standard Gaussian distribution

See figure 4.3d. Now the table in appendix B does not show negative values of Z, but the symmetry of the Gaussian means that

$$\text{Prob} [-2.25 \le Z \le -1.75] = \text{Prob} [1.75 \le Z \le 2.25]$$
$$= \text{Prob} [Z \le 2.25] - \text{Prob} [Z \le 1.75]$$
$$= 0.988 - 0.960 = 0.028.$$

Thus there is less than a 3 percent chance that the shopping center will be built for between 70 percent and 80 percent of its estimated cost. (This example is as tricky as these Gaussian problems get, so if you feel comfortable with it, you are in good shape. Be sure you understand the step

$$\text{Prob} [1.75 \le Z \le 2.25] = \text{Prob} [Z \le 2.25] - \text{Prob} [Z \le 1.75].)$$

Summary

The hypergeometric distribution describes the composition of samples taken from finite populations comprised of two kinds of cases. The binomial distribution describes the number of successes in a sequence of identical Bernoulli trials. A Bernoulli trial is an independent random event with only two possible outcomes. The beta distribution describes a continuous random variable confined between zero and unity, such as an unknown probability. The Poisson distribution describes the random count of events in time or space, and can be thought of as the limiting case of a binomial distribution with a large number of trials and the same small probability of success on any single trial. The Gaussian distribution is commonly used to describe continuous random variables, especially sums.

References and Readings

Bartlett, M. *The Statistical Analysis of Spatial Pattern*. London: Chapman and Hall, 1975.

Bishop, Y. M. M., S. E. Feinberg, and P. W. Holland. "Sampling Models for Discrete Data." *Discrete Multivariate Analysis*. Cambridge, Mass.: The MIT Press, 1976, pp. 435–456.

Ord, J. K. "Graphical Methods for a Class of Discrete Distributions." *Journal of the Royal Statistical Society* (A) 130 (1967): 232–238.

Tukey. "Shapes of Distributions," chapter 19, pp. 615–645.

Ripley, B. "Modelling Spatial Patterns" (with Discussion). *Journal of the Royal Statistical Society* (B) 39 (1977): 172–212.

4.1

An ad-hoc board is to be established to oversee open space issues in an exurban community. Board members are to be selected by lot from among 10 candidates, of whom 4 are pro-growth and 6 are zero-growth advocates. At issue is whether the board should have 3 or 5 members. Which size would be more likely to produce a pro-growth majority?

4.2

Two housing rehabilitation programs are to be compared. Each dwelling unit will be classified as successfully rehabilitated or not, according to criteria relating to physical condition of the unit and time and cost of the rehab effort. In reality, program A can successfully rehab a unit with probability 0.9, whereas the probability is only 0.5 with program B. A small pilot study will compare the two programs by assigning each 4 units to rehab. What is the chance that the better program will successfully rehab a greater number of units?

4.3

Schnelle et al., reported the counts of serious crimes in a 1.65 square mile area of Nashville, Tenn. The counts were recorded during the 9 A.M. to 5 P.M. shift over 52 consecutive shifts. Compare the distribution of crimes to that predicted by a Poisson model.
1, 2, 3, 1, 1, 0, 0, 0, 1, 4, 1, 1, 0, 1, 0, 1, 1, 0, 2, 1, 0, 0, 0, 1, 1, 4, 0, 0, 0, 2, 2, 2, 2, 0, 1, 1, 1, 2, 0, 2, 0, 1, 2, 0, 2, 0, 2, 1, 1, 1, 1, 1.

4.4

Heckman and Willis studied the labor force participation of 1,583 married white women and concluded that the probability that a woman will work in any given year varies greatly from woman to woman. The study lead to a mover / stayer hypothesis which holds that some women have a very high probability of working in any given year while others have a very low probability. Heckman and Willis used the beta distribution to describe the distribution of the probability of working. Graph the beta distribution once with parameters $(S = 2, \ F = 1)$ and again with parameters $(S = -1/2, \ F = -1/2)$. Which graph corresponds to a mover / stayer hypothesis? (Note: the factorial can also be computed for noninteger arguments: $(-1/2)! = 1.772$.)

4.5

It has been observed that the number of occupied beds in a hospital maternity unit is often well described by a Poisson distribution. It is also known that the probability that a Poisson variable exceeds some value is often well approximated by the equivalent probability for a Gaussian variable with the same mean and standard deviation. Thus the right-hand tail of the Gaussian distribution can be used for estimating the number of maternity beds necessary to accommodate a given average demand. Suppose the number of occupied beds is Poisson with mean 15. Use the Gaussian approximation to estimate the number of beds necessary to accomodate 95 percent of the demand.

4.6

The distribution of family money income among all U.S. families in 1976 is given in the following table. Fit a Gaussian distribution to log income and assess the quality of the fit informally. (Hint: use the 30 percent point to estimate the standard deviation.)

Income class (in dollars)	Percentage of families
> 3,000	3.9
3,000–4,999	6.5
5,000–5,999	3.9
6,000–6,999	3.9
7,000–9,999	11.8
10,000–11,999	8.1
12,000–14,999	12.1
15,000–19,999	19.1
20,000–24,999	12.9
≥ 25,000	17.8
(Median = 14,958)	

II MAKING ESTIMATES

5 Estimating the Composition of a Finite Population by Sampling

A common prediction problem is to estimate from a sample of cases the number of a larger population that possess a certain attribute. For instance, we may be studying the 351 cities and towns of Massachusetts to determine how many provide municipal day care services. In each case, the truth will be a single number that we could determine exactly if only we could check every case. However, very often we can afford only to sample the population, in which case our response must be less definite, taking the form of a probability distribution over the range of possible answers. We will use Bayes' rule to combine the data from the sample with our prior expectations about the answer.

Drawing a Simple Random Sample from a Finite Population

Sampling theory provides predictions about a population based on what turns up in a sample of cases drawn from that population. The theory is valid only to the extent that the assumptions about how the sample was drawn are valid. If these assumptions are not true, then inferences based on the sample may very well be far off the mark. There are several kinds of sampling schemes, including simple random samples, stratified random samples, cluster samples, and systematic samples. We will consider only the most straightforward: *simple random sampling.*

A simple random sample is drawn from the population in such a way that every possible *pattern* of cases has an equal chance of constituting the sample. If the population consists of N cases and the sample of n cases, then there are $\binom{N}{n} = \dfrac{N!}{(N-n)!n!}$ possible samples of size n, and each must be equally likely. For instance, suppose there are 6 cities in a population and we wish to draw a simple random sample of 4 cities. There are $\binom{6}{4} = 15$ possible samples that can be drawn, and each must be equally likely. The 15 possibilities are shown in table 5.1. If certain combinations are less likely to occur—or impossible—for any reason (perhaps the mayors of cities A and B will never agree to participate in the same study), then the sample is not a simple random sample, and the usual statistical inferences will be in error to an unknown extent.

In practice, it is not necessary to put all $\binom{N}{n}$ possible samples in a hat for a drawing; this is fortunate, since, for example, there are $\binom{351}{50} = 1.6 \times 10^{61}$ possible samples of 50 Massachusetts cities and towns. Instead, we

Table 5.1
The 15 possible samples of size 4 from a population of 6 cases

Sample	City					
	1	2	3	4	5	6
1	X	X	X	X		
2	X	X	X		X	
3	X	X	X			X
4		X	X	X	X	
5		X	X	X		X
6			X	X	X	X
7	X		X	X	X	
8		X	X	X	X	
9	X			X	X	X
10		X		X	X	X
11			X	X	X	X
12	X	X			X	X
13	X		X		X	X
14	X			X	X	X
15	X	X	X			X

can assign each member of the population a number, find a list of n random numbers drawn from a uniform distribution, and choose the n cases corresponding to the n random numbers. In our example we want to draw a sample of 4 from 6 cities, which we number 1 through 6. We find the following list of uniform random numbers between 0 and 99 (see Schmitt, p. C1, line 28):

53, 53, 81, 82, 03, 08, 27, 77, 53, 46, 67, 85, 95, 06, 34, 56, 02.

We then scan the list in order, selecting a city if its number is the same as the first digit of the random number, skipping repeats and numbers 00 to 99. The steps would be

Step	Random number	Corresponding action
1	53	Select city 5
2	53	Skip it; city 5 already selected
3	81	Skip it; no city 8

Step	Random number	Corresponding action
4	83	Ditto
5	03	Skip it; no city 0
6	08	Ditto
7	27	Select city 2
8	77	Skip it; no city 7
9	53	Skip it; city 5 already selected
10	46	Select city 4
11	67	Select city 6
12	STOP—We now have a sample of 4 cities.	

This particular sequence of random numbers produced the sample consisting of cities 5, 2, 4, and 6. The process is mathematically equivalent to putting the $\binom{6}{4} = 15$ possible quartets on slips of paper into a hat and pulling one out at random. If we had wanted a sample of 50 cities from among the 351 in Massachusetts, we would have followed the same general procedure, this time using 3-digit random numbers (or the first 3 digits of larger random numbers) so that cities 100 through 351 would have a chance of being selected.

Bayes' Rule and the Prediction

Recall that our purpose is to infer, from a sample, how many members of a larger but finite population possess some particular attribute. Once we have randomly selected the cases, we must examine them in the sample to determine how many possess the attribute in question. Having done this (in practice this may not be easy, especially if we face the problem of non-cooperation or nonresponse), how do we make the prediction about the number with the attribute in the population? We leap from the sample (and also, if we wish, from our prior suspicions about the answer) to the population by means of Bayes' rule and the hypergeometric distribution.

Suppose the population contains $N = 100$ cases, of which an unknown number M possess the attribute, and the sample contains $n = 17$ cases, of which $m = 3$ possess the attribute. Then Bayes' rule can be written:

$$\text{Prob} \left[\begin{array}{c|c} M \text{ in population} & 3 \text{ in sample} \\ \text{of } 100 & \text{of } 17 \end{array} \right]$$

$$= \frac{\text{Prob} \begin{bmatrix} 3 \text{ in sample} & M \text{ in population} \\ \text{of } 17 & \text{of } 100 \end{bmatrix} \times \text{Prob} \begin{bmatrix} M \text{ in population} \\ \text{of } 100 \end{bmatrix}}{\sum_{k=0}^{100} \text{Prob} \begin{bmatrix} 3 \text{ in sample} & K \text{ in population} \\ \text{of } 17 & \text{of } 100 \end{bmatrix} \times \text{Prob} \begin{bmatrix} K \text{ in population} \\ \text{of } 100 \end{bmatrix}}.$$

Thus for any possible answer M we can calculate a conditional probability that expresses our relative confidence that M is indeed the answer rather than $M + 12$ or $M - 3$ or some other possibility. Since we really care only about the relative likelihood for each possible value of M, we can ignore the denominator on the right hand side, since it is only a normalizing constant. We are left with the proportional relationship

$$\text{Prob} \begin{bmatrix} M \text{ in population} & 3 \text{ in sample} \\ \text{of } 100 & \text{of } 17 \end{bmatrix}$$

$$\alpha \, \text{Prob} \begin{bmatrix} 3 \text{ in sample} & M \text{ in population} \\ \text{of } 17 & \text{of } 100 \end{bmatrix} \times \text{Prob} \begin{bmatrix} M \text{ in population} \\ \text{of } 100 \end{bmatrix}.$$

More generally, for a sample of size n, containing m cases with the attribute of interest, drawn from a population of size N containing an unknown number M of cases with the attribute

$$\text{Prob } [M \,|\, n, \text{ N}, m] \; \alpha \; \text{Prob } [m \,|\, n, \text{ N}, M] \times \text{Prob } [n, \text{ N}, M].$$

The term Prob $[n, \text{ N}, M]$ is called the *prior distribution* of the random variable M and summarizes our knowledge about the unknown M before the sample results are known. The term Prob $[m \,|\, n, \text{ N}, M]$ is called the *likelihood function* and is the probability that the sample results would be as they are, conditional on the value of M. (It is also conditional on n and N, but neither of these is unknown.) The term Prob $[M \,|\, n, \text{ N}, m]$ is called the *posterior distribution* of the random variable M and summarizes our knowledge about the unknown M after the sample results are available. The prior distribution might be based solely on professional judgment, on a mixture of subjective judgment and empirical data, or solely on data. The likelihood function arises from a probability model of the sampling process; in this case the likelihood function is the hypergeometric distribution introduced in chapter 4. The posterior distribution is a modification of the prior distribution that incorporates the empirical sample results. If we can plug into Bayes' rule numbers for the likelihood and prior probability, we can compute our relative confidence that the answer is some particular value of M. Then we can compare these relative probabilities for various values of M to see which choice of M is the best estimate and how much better it is than other possibilities.

The Likelihood Function

This probability can be derived from the theory of taking simple random samples from finite populations comprised of two types of cases. Given the population size N, the sample size n, and a hypothetical value for M, the chance of getting m cases with the attribute in the sample is the hypergeometric probability

$$\frac{\binom{M}{m}\binom{N-M}{n-m}}{\binom{N}{n}}.$$

For instance, if there really are $M = 14$ in a population of $N = 100$, then the chance of turning up $m = 3$ in a sample of $n = 17$ is

$$\frac{\binom{14}{3}\binom{86}{14}}{\binom{100}{17}} = \frac{(3.64 \times 10^2) \times (4.54 \times 10^{15})}{(6.65 \times 10^{18})} = 0.25.$$

The Prior Probability

The other piece of the prediction involves the prior probability, which embodies a judgment about the answer before taking the sample. The great value of Bayes' rule arises from its ability to combine in a simple, consistent way sample data and professional judgments (as expressed in the distribution of prior probabilities).

Where do prior probabilities come from? They may be completely empirical, summarizing the results of previous studies. For instance, suppose previous studies had found values of M/N of 0.13 in Baltimore, 0.18 in Columbus, 0.26 in Philadelphia, and 0.42 in South Hadley. Then you might use the prior distribution in figure 5.1a for your population of $N =$

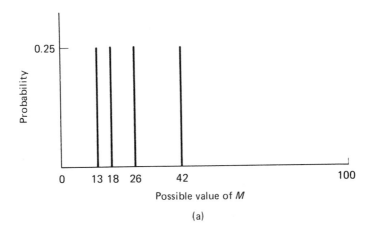

Possible value of *M*

(a)

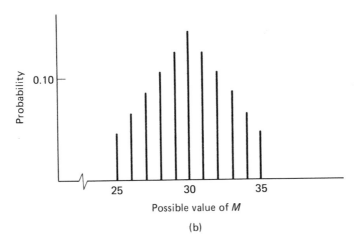

Possible value of *M*

(b)

Figure 5.1
Two possible prior distributions

100 cases. Alternatively, the prior might be determined more subjectively. You might feel in your bones that the true value of M could not be less than 25 nor greater than 35, with 30 most likely, and posit the prior shown in figure 5.1b. This kind of latitude bothers those with "scientific" habits because it seems too uncontrolled and subjective, but its appeal should be clear to planners, whose judgments often rely as heavily on expert opinion and intuition as on data. Besides, it is commonly noted in the planning world that two people will react quite differently to the very same data, and the Bayesian approach explains this basic phenomenon as a difference in prior distributions leading to a difference in posterior distributions.

There is one particular prior distribution that deserves special mention. If you wish to make the weakest possible prior assumption about the answer, you should chose the prior distribution with the greatest entropy, meaning that uncertainty about the answer is as great as can be. For the present problem, this distribution is the discrete uniform distribution that specifies that all $N + 1$ possible values of M between 0 and N are equally likely, each having a prior probability $\dfrac{1}{N + 1}$.

The Posterior Probability

Having both the prior and the likelihood, we can put the pieces together to finally make the estimate—or, more properly, since we are quite conscious of the uncertainty inherent in generalizing from a sample result, to get the relative likelihood of the various possible estimates. Each possible answer M has a relative credibility given by Bayes' rule:

$$\text{Prob}\,[M\,|\,n,\,N,\,m] \;\alpha\; \frac{\dbinom{M}{m}\dbinom{N - M}{n - m}}{\dbinom{N}{n}} \times \text{Prob}\,[n,\,N,\,M].$$

Example: Detoxification Centers with Long and Short Stays

Consider a specific example. There were 22 alcoholic detoxification centers in Massachusetts in 1975. We wish to know how many of the 22 had an average length of stay of more than 3 days. Since most detox centers serve about 2,000 or more clients each year, a record review to determine average length of stay at each facility would take a long time. (In practice we might sample the records as well as the centers, but let us assume we know with certainty whether the length of stay exceeded 3 days in each center sampled.) Suppose we will pick a random sample of 10 of the 22 detox centers and make an estimate based on this sample. What predictions will follow? How close will they be to the truth?

We begin with the prior distribution. We know the possibilities range from 0 to 22. But which numbers are more likely? Having no previous experience with detox centers, we fall back on the assumption of maximum ignorance and assume all of the 23 possible values of M to be equally likely a priori, each having prior probability $1/23$. Now only the likelihood function will influence the posterior distribution.

We need a sample. Following the random selection procedure outlined earlier, we select one from among the $\binom{22}{10} = 646{,}646$ possible samples of size $n = 10$ and then determine the average length of stay in each center. The sample results are listed in table 5.2. How do we get the posterior probabilities? We know immediately that there can be no fewer than 6 and no more than $22 - 4 = 18$ detox centers with average stay greater

Table 5.2
Sample data for detoxification center example[a]

Detox center chosen	Average stay exceeding 3 days
10	No
9	Yes
1	No
17	Yes
20	No
5	No
4	Yes
2	Yes
16	Yes
8	Yes

Source: Massachusetts Department of Public Health, *1976 Health Data Annual*, pp. 136–145.
[a] Since 6 of the sample of 10 had average stays exceeding 3 days, a good quick guess is that 6/10 of the total of 22 centers, that is, 13.2 centers, have average stays in excess of 3 days.

than 3 days since we found 6 cases with the attribute and 4 without in the sample. For the possible answers $6 \leq M \leq 18$ we compute

Prob [M in 22 | 6 in 10] α Prob [6 in 10 | M in 22] $\times$ Prob [M in 22].

Using the hypergeometric likelihood and the uniform prior, we have

$$\text{Prob } [M \text{ in } 22 | 6 \text{ in } 10] \ \alpha \ \frac{\binom{M}{6}\binom{22-M}{4}}{\binom{22}{10}} \times \left(\frac{1}{23}\right),$$

and ignoring constant terms that carry no real information

$$\text{Prob } [M \text{ in } 22 | 6 \text{ in } 10] \ \alpha \ \binom{M}{6}\binom{22-M}{4}.$$

The prior and posterior distributions of M are shown in figure 5.2. The sample has moved us from our prior distribution of $1/23$ probability for each possible value to a more definite posterior that gives no credibility to answers less than 6 or more than 18 and gives most credibility to a prediction that 13 of the 22 detox centers have average stay in excess of 3 days, although $M = 14$ is very nearly as likely. The entropy of the prior was

$$E = -\sum_{i=0}^{22} P_i \log P_i = -\sum_{i=0}^{22} \frac{1}{23} \log \frac{1}{23} = -23 \left(\frac{1}{23} \log \frac{1}{23}\right)$$

$$= -\log \frac{1}{23} = \log 23 = 1.36,$$

while the entropy of the posterior is 0.97, so the sample reduced our uncertainty by 29 percent.

What would have happened if we had used a different prior? Suppose we unearthed a previous study of 10 Massachusetts detox centers (not otherwise identified to protect their anonymity) which noted in passing their average lengths of stay, all of which exceeded 3 days. Suppose that we also know that since the earlier study was completed there has been a general trend toward longer stays. We might put this information together to formulate a prior that the answer is somewhere between 10 and 22 but certainly no fewer than 10. So we might say that each of the 13 alternatives between 10 and 22 inclusive has a prior probability of $1/13$. This more informative prior distribution and its posterior distribution are shown in

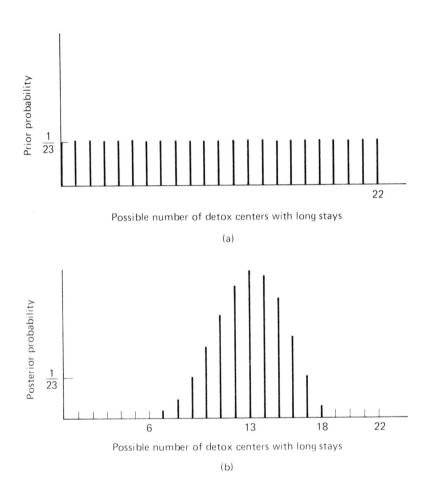

Figure 5.2
A flat prior and its posterior distribution for the detox center example

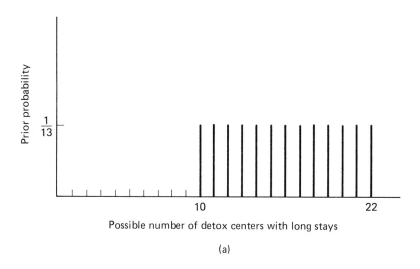

(a)

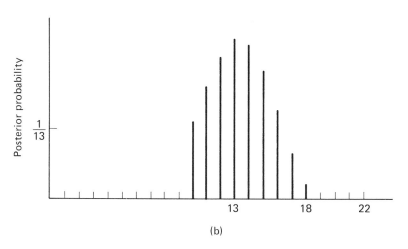

(b)

Figure 5.3
A more informative prior and its posterior distribution for the detox center example

figure 5.3. Our best guess still happens to be that there are 13 detox centers with average stays exceeding 3 days, although again 14 is very nearly as good an answer as 13. Our more definite prior, however, has cut down on the range of answers that have any credibility. The sample has transformed our more definite prior with entropy 1.11 to a posterior with entropy 0.89, a 20 percent reduction.

What is the true answer? In this case we know the answer because the Massachusetts Department of Public Health has collected and reported the data from all 22 centers (*1976 Health Data Annual*, pp. 136–145). The true answer happens to be 14 of 22. Did we do well or poorly? Our best estimate, 13, was not correct but is close. Furthermore, the true answer, 14, had nearly as high a probability in the posterior distribution as did 13, so we are not very surprised to learn that the true answer is 14. I think we did rather well: we got a good fix on the answer by studying fewer than half the detox centers. In practice we would not usually have access to the true answer until much later than we need it, if ever, so we could not second-guess the sampling process this way. Instead, we would have to rely on the sampling process and the Bayesian inference procedure to both get us close to the truth and keep track of the uncertainty arising from examining only some of the cases.

Could we not have gotten the same prediction much more easily by just noting that $(6/10) \times (22) = 13.2$ and immediately using 13 as our estimate? Of course, and that calculation would help us get zeroed in. Then what is the payoff to all the arithmetic associated with calculation of the posterior? The payoff is that we learned from the posterior how "hard" our estimate was, what range of values might contain the true answer. We learned that 13 was barely superior to 14 as an estimate, and that the sample was only sufficient to narrow our focus to a range of perhaps 11 through 14. The posterior distribution displayed to us the uncertainty about our estimate. The Bayesian analysis also gave us a way to systematically include any prior knowledge we had about the answer. The commissioner of public health may insist on just a simple 13 as an estimate, but as the commissioner's advisor you would have the responsibility to keep track of how solid such an estimate really is. And if the number were particularly crucial to some program, even the commisioner may insist on your producing a range of estimates.

Two interesting questions remain and may have occurred to you. First,

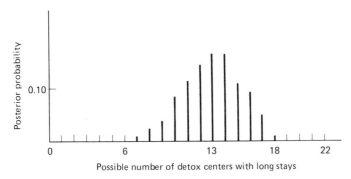

(a) Sample result: 6 of 10 have long stays

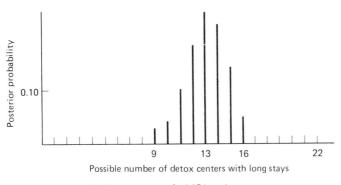

(b) Sample result: 9 of 15 have long stays

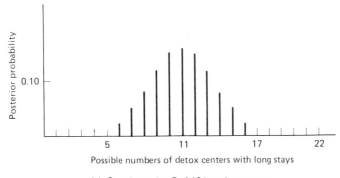

(c) Sample result: 5 of 10 have long stays

Figure 5.4
Posterior distributions from flat prior and three different sample results

since there are $\binom{22}{10} = 646{,}646$ possible samples of size 10 from among the 22 detox centers, how might the results have been different if we had drawn one of the other 646,645 possible samples? Second, how much would the estimate have been improved by taking a larger sample? I will not provide exhaustive answers to these questions, but to help you develop a feel for the kinds of variation that arise from the vagaries of sampling and for the utility of larger samples, we display for comparison in figure 5.4 the posterior distributions arising from flat priors and (a) our original sample of 10 detox centers, which found 6 of 10 with long stays, (b) the original sample of 10 augmented by an additional 5 which uncovered 3 long-stay and 2 short-stay centers, for a total of $6 + 3 = 9$ of 15 with average stays greater than 3 days, and (c) a new sample of 10, split 5 and 5. The sample (b) of 15 centers also happens to lead to an estimate of 13, but with less uncertainty due to the larger sample size. The new sample of 10, which split 5 : 5, gives 11 as the best estimate, which is a poorer estimate than that produced by the first sample. Take some comfort, though, from the fact that the 5 : 5 split will occur only half as often in a sample of 10 as a 6 : 4 split, since

$$\frac{\binom{14}{5}\binom{8}{5}}{\binom{22}{10}} = 0.17 \quad \text{while} \quad \frac{\binom{14}{6}\binom{8}{4}}{\binom{22}{10}} = 0.33.$$

Summary

A common problem is to determine the number of cases in a population of finite size that possess a particular attribute. A simple random sample can provide an estimate of the number. The Bayesian approach to the problem combines prior knowledge or judgment about the relative probabilities of the possible answers with the empirical results of the sample to produce a posterior distribution of relative probabilities. The likelihood function for such problems is a hypergeometric probability. The simplest form of the computations is

Prob $[M$ in population of $N \,|\, m$ in sample of $n] \,\alpha\, \binom{M}{m}\binom{N-M}{n-m}$

$\times$ Prob $[M$ in population of $N]$.

References and Readings

Jeffreys, H. "Estimation Problems." *Theory of Probability*, 3d. ed. Oxford: Clarendon Press, 1961, pp. 99–167.

Sedransk, J., and J. Meyer. "Confidence Intervals for the Quantiles of a Finite Population: Simple Random and Stratified Random Sampling." *Journal of the Royal Statistical Society* (B) 40 (1978): 239–252.

Problems

5.1

Suppose you are responsible for administrative oversight of seven neighborhood service centers. Word reaches you that a team of investigative reporters is working on a story that will charge financial impropriety at one center and cite rumors and indications of impropriety at others. You know that the newspaper is usually able to document its direct claims of public wrongdoing but also tends to overplay its softer information. Your general sense of the program is that corruption is possible but not likely, and a recent random audit of one center indicated no problems. You are able to order and complete two additional audits in randomly chosen centers before the HEW regional coordinator summons you to a meeting about the impending crisis. What would you have reported to the HEW official about the extent of the scandal, if any, prior to the two additional audits? What would you report after the audits?

5.2

Seligman made an intensive study of the finances of the 46 United States in the late 1880s. State your personal prior distribution for the number of states that were debt-free at the time. A simple random sample of 17 states reveals that m were debt-free, where m is the solution to the equation $m^2/2 = \sqrt{1024}$. (The sample result is encoded thus to help you keep your prior truly prior.) Determine the likelihood and posterior distributions.

5.3

A housing advocacy group desires to review the performance of a newly appointed zoning board of appeals. To date the board has heard 10 appeals, granting 6 and denying 4. The advocacy group suspects that the board is biased in favor of applicants representing large-scale developments, so the group plans to crosstabulate the board's decisions and the scale of the proposed developments as follows:

		Decision		
		Approve	Deny	
Scale	Largest			5
	Smallest			5
		4	6	10

a. List all the possible combinations of counts that might result from the study. Determine the probability of each combination under the assumption that the board takes no account of scale when making its decisions.
b. If the advocacy group regards a higher percentage of approvals for large-scale projects as evidence of bias, what is the chance that the group would wrongly accuse a board that ignores scale?

6 Estimating a Probability

Another commonly encountered estimation problem for planners amounts to estimating the probability of success in a Bernoulli trial. Estimates include:

• What fraction of high-risk infants will survive in a regional system providing neonatal care?
• What is the probability that an emission source will violate air quality standards under a new enforcement system?
• What is the chance that a site will be developed within five years if a zoning variance is granted for its development?
• What is the probability that a delinquent youth receiving special prerelease services will be rearrested within one year?
• What is the likelihood that a landlord will cease maintaining his units if rentcontrol is enacted?

How do these problems differ from counting the number of cases in a finite population that possess a certain attribute (after all, the count can also be expressed as a proportion of the population)? First, there need not be any preidentified, finite population; for instance, the question about neonatal mortality is better thought of in terms of a series of coin flips than as a sampling from a population of babies not yet conceived. In these problems the attribute is viewed prospectively. Second, the answer is a continuous random variable, not discrete: whereas in a population of 23 cases, the proportion possessing some attribute must be a multiple of $1/23$, the probability that a site will be developed might be any number between zero and unity. Third, although the Bayesian approach applies to both estimation problems, the calculations are somewhat easier for the problem of estimating a probability, since the likelihood function is the binomial probability distribution not the hypergeometric.

The paradigm of coin flipping applies to the estimation of a probability. We will assume we have data on the outcomes of a number of cases and that each case represents a Bernoulli trial:

1. Each case will be either a success or a failure.
2. All cases are identical in that they have the same underlying probability of success (which we seek to estimate).
3. The outcome of each case is independent of the outcomes of all the other cases.

Assumption 1 seems straightforward enough but does require that we have an unambiguous criterion for success. Assumption 2 means we are ignoring the attributes that distinguish one case from another—we are dealing with average cases, although we can minimize the variability among cases by restricting the data cases to those considered most pertinent (for example, consider as data not all recent cases of proposed site development but only those involving the developer in question). Assumption 3 is an independence assumption that holds the golden key to analytical tractability but must always be carefully scrutinized. It may be a bad assumption, for instance, when looking at the probability of an owner selling a home in a changing neighborhood; it should be OK when predicting rearrest probabilities for juveniles. The planner must develop a touch for cutting the Gordian knot of dependence, controlling both the urge to blindly assume independence and get on with the analysis and the urge to meditate forever on the mystical oneness of all phenomena.

The Bayesian Estimate

Let us review the circumstances of the problem. What do we want to know? The chance that a success (or failure) will occur. What form do we wish the answer to take? A probability distribution expressing our relative degree of confidence in each of the possible answers between 0 and 1.0. What do we already know about the answer? We bring to bear a combination of our prior expectations about the answer (which are perhaps based on empirical knowledge) and a set of sample data. What is the set of data? A sample of trials in which the outcome of each case was recorded as a success or a failure. We assume that the trials were independent and that the cases were sufficiently alike that they share a common probability of success (which is the unknown we seek to estimate).

As before, we use Bayes' rule to combine properly both our prior feelings about the answer and the results of our sample of cases. The true answer is some number between zero and one. All numbers within that range are theoretically possible answers, but some are much more likely to be the answer than others because they are more consistent with both our prior judgment and the sample results. We express our knowledge about the answer in the form of the posterior distribution. A handy form of Bayes's rule for this problem is

Posterior [probability | data] α Likelihood [data | probability]
  $\times$ Prior [probability].

A final note on how your estimate relates to prediction. Suppose you are trying to estimate the probability that a new town will be financially viable. By reviewing the experience of six similar new towns you obtain sample data, which you merge with a prior distribution based on your knowledge of the particular developer. You compute a posterior distribution that turns out to be fairly narrow and centered around 0.8. You decide that this high probability of success is sufficient to warrant an investment in the town. You invest; if the town fails, you wonder what went wrong. You must remember that the outcome of any single trial will be either a success or a failure. All the estimate does is tell you which outcome will be more surprising. In the case of the new town development, it might be that exactly 80 of 100 such ventures would succeed. You, however, are dealing with but one case and are therefore more vulnerable to surprise. It still makes sense to use the analytical techniques to organize all the information available; the explicit consideration of uncertainty should sharpen and focus your perception of risk. And the width of the posterior distribution will be a graphic representation of your level of confidence about the predicted probability of success.

The Likelihood Function

As always the likelihood function expresses the probability of obtaining the observed sample results conditional on an hypothesized value of the random variable to be estimated. In this case the variable to be estimated is the probability of success in a Bernoulli trial, which we will call P. The sample results are the total number of successes and failures in the sample, which we will call S and F, respectively. As we noted in chapter 4, the distribution describing a collection of Bernoulli trials is the binomial. Thus

$$\text{Likelihood } [S, F \,|\, P] = \binom{S + F}{S} P^{S}(1 - P)^{F}.$$

Suppose, for example, that we wish to estimate the probability P that a factory will reduce its emission of pollutants under a new system of clean air incentives. A sample of 6 factories shows that $S = 4$ reduced their emissions and $F = 2$ did not. Thus

$$\text{Likelihood } [S = 4, F = 2 \,|\, P] = \binom{4 + 2}{4} P^4 (1 - P)^2$$
$$= 15P^4 (1 - P)^2.$$

Now if we hypothesized that the true value $P = 0.1$, then the probability of obtaining 4 successes in 6 trials would be only about 0.001. On the other hand, if we hypothesized $P = 0.9$, then the chance of obtaining the sample results actually observed would be about 80 times greater, so the hypothesis $P = 0.9$ is much more consistent with the actual sample data than is the hypothesis $P = 0.1$. Of course we would expect $P = 4/6$ to be most consistent with the data, and in fact the value of the likelihood function in this case is about 0.33, which is indeed greater than the likelihood for any other value. Since the important part of the binomial distribution is $P^S (1 - P)^F$, a graph of the likelihood function has the same shape as a graph of the beta distribution (see figure 4.1).

The Prior Distribution

The prior distribution is a curve that expresses our relative belief in the possible values of P before we take the sample and obtain data. Of course this curve might be of any shape, depending on our prior sense of optimism or pessimism about the chance of success and on the confidence we attach to that prior sense. One particularly interesting prior, which expresses maximum ignorance about the value of P, is the uniform distribution. If we want to inject no prior biases, relying only on the sample results to determine the posterior distribution, or if we really have no good idea about what value of P to predict, then we use the uniform distribution as our prior.

In general, the estimation procedure is to decide on a prior distribution, take a sample of data, compute the likelihood function, then multiply the prior by the likelihood to obtain the posterior distribution (technically, we should insure that the area under the posterior is unity so we will have a legitimate probability distribution, but in practice we may care only to know the relative shape of the posterior distribution). If a computer were multiplying the prior by the likelihood for you, generating the posterior distribution would not be difficult. However, if you were doing the multiplications by hand, the task would be tedious if the prior distribution were some arbitrary curve. Luckily, there is a way to solve these estimation problems with minimal computation.

The hero is the beta distribution. It has a useful property of regeneration: a beta prior distribution, when multiplied by the binomial likelihood function, emerges as another beta distribution for the posterior. Furthermore, the beta has just two parameters, which can be chosen to produce a wide range of shapes, so that many useful prior distributions (including the uniform) can be expressed as beta distributions. Although there is a normalizing coefficient in the beta distribution, the important part of the beta is $P^S(1 - P)^F$. You will recognize this also as the essence of the binomial likelihood function. What this means is that we can conceive of our prior distribution as if it had arisen from a given set of empirical successes and failures (as it may well have), obtain sample data, and then express the posterior in the same terms. Letting S_{prior} and F_{prior} be the two parameters of the beta prior distribution, we can express the belief that the probability of success, P, is high by making S_{prior} bigger than F_{prior}; for instance, $S_{\text{prior}} = 3, F_{\text{prior}} = 0$ will make values of P near the mode $S/(S + F) = 3/(3 + 0) = 1.0$ more likely a priori than lower values. We can also control the degree of confidence expressed in the prior by controlling the sum $(S_{\text{prior}} + F_{\text{prior}})$: larger sums lead to "skinnier" prior distributions. Thus, if we feel that the probability of success is a bit better than 0.5, but we are not very confident about this, we can pick S_{prior} and F_{prior} in such a way that the ratio $S_{\text{prior}}/(S_{\text{prior}} + F_{\text{prior}})$ is slightly greater than 0.5 and the sum $S_{\text{prior}} + F_{\text{prior}}$ is not too large. For instance, we might choose $S_{\text{prior}} = 6$, $F_{\text{prior}} = 4$, which would give the moderately and cautiously optimistic prior shown in figure 6.1.

The Posterior Distribution

If we chose the convenience of a beta prior distribution, we can move quite simply to the beta posterior distribution, whose parameters are computed by addition:

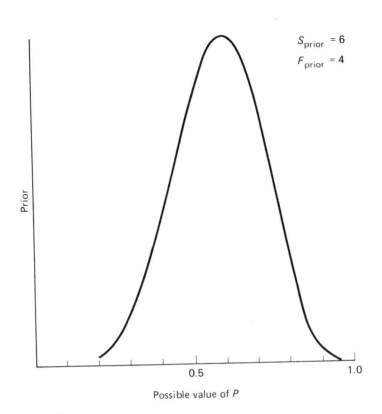

Figure 6.1
A beta prior distribution

$$S_{post} = S_{prior} + S_{data}$$
$$F_{post} = S_{prior} + F_{data}.$$

This is because

Posterior $[P|\text{data}] \propto$ Likelihood $[\text{data}|P] \times$ Prior$[P]$

becomes

Posterior $[P|\text{data}] \propto P^{S_{data}}(1 - P)^{F_{data}} \times P^{S_{prior}}(1 - P)^{F_{prior}}$
$$\propto P^{S_{data}+S_{prior}}(1 - P)^{F_{data}+F_{prior}}$$
$$\propto P^{S_{post}}(1 - P)^{F_{post}}.$$

Suppose in our example about air quality controls we felt somewhat confident a priori that the chance of success was about 0.6. We might therefore have chosen a beta prior with $S_{prior} = 6$, $F_{prior} = 4$. Our experimental trials produced the results $S_{data} = 4$, $F_{data} = 2$. Thus the posterior mode occurs at $S/(S + F) = 10/16 = 0.63$. Note how the sample data have transformed the prior distribution. The sample results have shifted our estimate of the probability of success upward a bit; they have also improved our confidence in the estimate by narrowing the distribution.

In this case, we expected from our prior distribution that about $6/(6 + 4) = 0.60$ of the trials would be successes. The data were relatively consistent with this prior distribution, since $4/6 = 0.67$ were successes. What if we had a prior distribution that predicted results very different from those obtained in the sample?

In such a situation, we get a more dramatic view of the way sample results combine with prior distributions. Suppose we had a very pessimistic prior, corresponding, say, to a beta distribution with parameters $S_{prior} = 0$, $F_{prior} = 6$. The sample results $S_{data} = 4$, $F_{data} = 2$ would lead to a beta posterior with parameters $S_{post} = 0 + 4 = 4$, $F_{post} = 6 + 2 = 8$. See figure 6.2. In this case the rather positive sample results force an upward adjustment in the estimate of the probability of success, although not as far as the sample data alone would indicate. In effect, the prior is treated like a previous set of six trials that produced no successes; indeed, the prior might in fact have been generated by six earlier trials. Note that the sample data have in this case increased the uncertainty in the estimate, which seems proper since the sample results were so surprising in relation to the pessimistic prior.

What if we had started with a very optimistic prior: say a beta distribution with parameters $S_{prior} = 6$, $F_{prior} = 0$? We would have arrived at yet another beta posterior distribution; this one with parameters $S_{post} = 10$, $F_{post} = 2$, as shown in figure 6.3.

Comparing the posterior distributions, we note that the same sample data (4 successes and 2 failures among factories subject to the air quality incentives) leaves those skeptical of the value of the incentives to remain skeptical, although less so, while those with a prior optimism remain optimistic, although less so. Is this odd? No: it is commonplace in practice for program advocates and program critics to give different interpretations

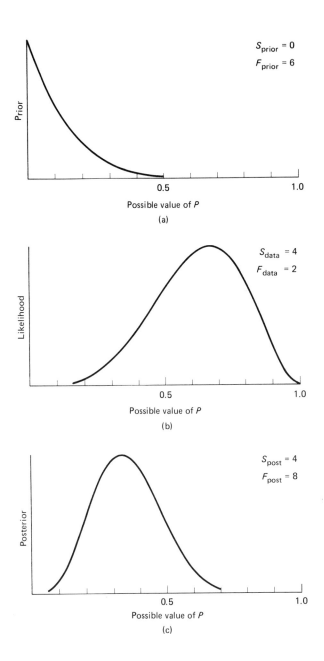

Figure 6.2
Estimation beginning with a pessimistic prior

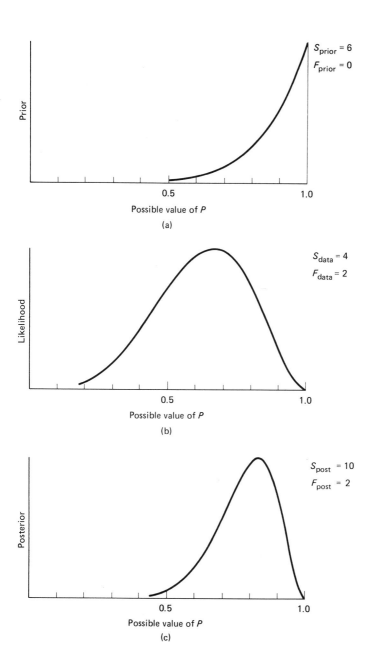

Figure 6.3
Estimation beginning with an optimistic prior

to the same data. The Bayesian estimation process allows for the consistent, orderly revision of opinion based on sample data. It also draws the two camps closer together—their posterior distributions are more alike than their prior distributions. If the sampling process were to continue so that large numbers of cases were studied, the empirical results would swamp the prior distributions; we would have learned so much about the air quality incentives that we would not need to rely so heavily on prior (perhaps subjective) estimates of the probability of success. Conversely, when we have very little data, our prior distributions play dominant roles. You can view this process of updating priors with sample results as a good way to approach the question, How much data do I need to convince a (rational) skeptic?

Highest Density Regions

The posterior distribution merges prior knowledge with sample data into an indication of the relative credibility of each possible value of the unknown probability of success. It provides a full picture of the uncertainty involved in making an estimate. However, on occasion we may chose to cite not the entire posterior distribution but rather two limits that are very likely to bracket the unknown probability. The set of possible values lying between these limits is called a *highest density region* (*HDR*). Graphically, a highest density region is determined as in figure 6.4. A horizontal line is lowered through the distribution, dividing it into 3 sections, until the central section contains some specified proportion of the area under the curve. For instance, the 95 percent *HDR* for the posterior distribution in figure 6.4 extends from roughly $P = 0.58$ up to $P = 0.97$, meaning Prob $[0.58 \leq P \leq 0.97] = 0.95$. Constructing the highest density region this way insures that every possible value within the *HDR* is more likely to be the true answer than any value outside the *HDR*. Quoting an *HDR* is intermediate between citing a measure of central tendency, such as the posterior mode or mean, and citing the entire posterior distribution: it conveys some, but not all, of the uncertainty associated with the estimate. Tables of *HDR*'s for the beta distribution are available in Schmitt's book.

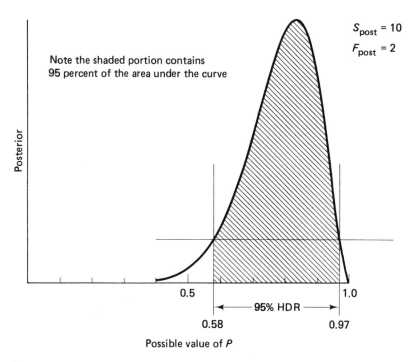

Figure 6.4
Finding the 95 percent highest density region (*HDR*)

Summary

A common problem is to estimate the probability of success P in a Bernoulli trial. The Bayesian approach to the problem determines a posterior distribution for P by multiplying a prior distribution by a binomial likelihood function based on sample totals of successes and failures. If the prior distribution is beta, then so is the posterior, which takes the form

Posterior $[P \,|\, \text{data}] \propto P^{S_{\text{post}}} (1 - P)^{F_{\text{post}}}$,

where

$S_{\text{post}} = S_{\text{prior}} + F_{\text{data}}$
$F_{\text{post}} = F_{\text{prior}} + F_{\text{data}}.$

Planners with different prior distributions will react differently to the same sample data, although their posterior distributions will be more similar than their priors, and, as sample size increases and evidence accumulates, their posterior distributions should converge. The uncertainty in the estimate of P can be expressed more succinctly but less completely by citing a highest density region rather than the full posterior distribution.

References and Readings

Schmitt. "The Rate Problem Revisited," section 4.3, and "Estimation of Rates," section 4.4, pp. 114–121, 121–129.

6.1

Okonjo studied indigenous capital markets in rural Nigeria, comparing use of the informal system to use of government-sponsored institutions like the post office savings bank. Consider the probability that a randomly chosen man from the town of Ogwashi-Uku has used the post office savings bank at least once. (a) What is your personal prior distribution (specify the values of S_{prior} and F_{prior}, and briefly explain your choice)? Graph your prior. (b) In a random sample of 50 men in Ogwashi-Uku, Okonjo found that S_{data} men had used the post office savings bank (so as not to influence your prior distribution, I have encoded the value of S_{data} in the formula

$$\sqrt{432/S_{data}^3} = 1.41421;$$

solve for S_{data} to learn Okonjo's finding). Graph the likelihood function and your posterior distribution.

6.2

Okonjo also studied the rate of loan defaults in indigenous credit associations. She found that in 1975 the Onyeliyachei Association suffered 2 defaults in 16 loans and in 1977 1 default in 9 loans. Regarding the 1975 results as the basis for a prior distribution and the 1977 results as subsequent sample data, graph the prior and posterior distributions of the default rate.

6.3

Using the Gaussian distribution as an approximation to the beta, determine an approximate 90 percent *HDR* for the default rate in the Onyeliyachei Association.

6.4

Branch and Fowler, in a community survey commissioned by the Massachusetts Department of Public Health, determined that among 169 elderly and chronically disabled individuals not residing in institutions only one required institutional care in a skilled nursing facility.

a. Branch and Fowler stated:

These numbers should be treated very cautiously. Although the rates are the best estimates available, small changes can make marked differences in the estimate of the number of people requiring additional services.

Assuming a flat prior and a total population of elderly and chronically disabled individuals of 700,000, depict graphically the uncertainty in the number of people living in the community who should instead be receiving skilled nursing facility care.

b. The Department of Public Health's Long Term Care Task Force, charged with determining the number of nursing home beds needed in Massachusetts, had this reaction to the survey finding:

...[The survey] failed to identify a statistically reliable unmet need for institutional services among the noninstitutionalized population ... any projection based on only one positive finding is statistically unreliable ... Since no unmet need among the noninstitutionalized population was demonstrated, ... [the bed planning methodology to be used] is essentially a reallocation of the currently institutionalized population.

Comment on the department's reaction in light of the fact that at the time of the survey there were about 15,000 skilled nursing facility beds in Massachusetts.

7 Estimating a Mean

The final estimation problem we will consider is that of estimating the mean value of a random variable, such as the mean number of neighborhood health center visits from each zip code, or the mean full-value tax rate among Massachusetts towns. Again we will assume that a simple random sample is drawn from a larger population whose mean we wish to estimate. To avoid certain mathematical complications we will further assume that the sample size n is very small relative to the population size N. (But please do not infer from this that a large sample, when you can get it, is less desirable than a smaller one.) Finally, we will assume that the random variable of interest, say, the full-value tax rate, has a Gaussian distribution in the population; later we will learn how to check this assumption and see how crucial it might be to our analysis.

The Sample Mean as an Estimate of the Population Mean

Suppose we wish to estimate the mean full-value tax rate among Massachusetts cities and towns (the full-value rate is the rate the city would be charging if all property were assessed at full market value). We could use the technique for drawing a simple random sample to select a handful of cases, compute the mean of the full-value tax rate for the sample cases—the *sample mean*—and make the approximation

Sample mean $\approx$ population mean.

For instance, I drew the simple random sample of $n = 11$ towns shown in table 7.1. Having drawn this sample, we could simply say that the average full-value tax rate in Massachusetts is about $27.42 and stop. Better yet, we could proceed to analyze the uncertainty inherent in our estimate, to learn how hard the estimate is. We have already seen that in the sample of 11 towns the tax rate ranges from $10.00 to $37.38; this range alone should give us pause and motivate a more careful treatment of the vagaries of sampling error. After all, if we had selected 11 other towns, we would almost certainly have gotten a different estimate. The beauty of sampling theory is that it allows us to parlay the variation observed within the sample into an expression of our uncertainty about the mean of the entire population.

We can proceed with a Bayesian analysis as before, treating the unknown population mean as a random variable whose various possible

Table 7.1
Sample data for tax rate example

Community	Full-value tax rate (1975)
Topsfield	$31.59
Northborough	31.36
Middleton	28.72
Nahant	31.08
Leicester	24.36
Savoy	13.86
Easthampton	37.38
Tyngsborough	33.12
Monterey	10.00
Wellesley	34.94
Belchertown	25.32
Sample mean	$27.42
Sample standard deviation	$ 8.59

Source: Massachusetts Taxpayers Foundation, Inc., "Municipal Financial Data, Including 1977 Tax Rates."

values have more or less credibility in light of our prior judgments and sample results. To simplify, we will again assume a flat (that is, uniform) prior distribution, so that the posterior distribution of the population mean will be proportional to the likelihood function alone. In a real problem you would be free to use an arbitrary prior distribution if you were willing to do some extra multiplications to compute the posterior (since posterior α likelihood × prior). As usual, you are not free to choose any likelihood function you want, since the likelihood depends on the probability model of the sampling process and the assumption of a Gaussian distribution in the population.

An Approximation to the Likelihood Function for Large Samples

We began with a small sample of values of the tax rate, wishing to produce a curve showing the likelihood that the particular sample mean we computed would arise if the tax rate had a Gaussian distribution in the population with true mean μ. Before we address this problem directly, we will ease in by considering the simpler problem of working with large samples, where "large" might mean samples of size $n \geq 50$ or so. The advantage of a large sample is that it will provide such a good estimate of the population standard deviation σ that we can use a Gaussian likelihood function. The reasoning is as follows:

The sample mean, which we denote by $\bar{x}$ (read "x bar"), is computed as the arithmetic average of the sample values:

$$\bar{x} = \frac{1}{n} \sum_{i=1}^{n} x_i.$$

Recall the statement in chapter 4 that sums of random variables tend to have approximately a Gaussian distribution. If the random variables being added are Gaussian to begin with, their sum is exactly Gaussian. And since the sample mean is just a scaled sum (the scaling factor is $1/n$), the mean of a sample drawn from a Gaussian population must itself be a Gaussian random variable. The value $\bar{x}$ will vary depending on which n cases constitute the sample, but we know that its distribution will be Gaussian. To specify a Gaussian distribution completely we need only provide its mean and standard deviation. It happens that if the population is Gaussian with mean μ and standard deviation σ, then the sample mean $\bar{x}$ is Gaussian with mean μ and standard deviation $\sigma/\sqrt{n}$, where n is the sample size.

The fact that the sample mean $\bar{x}$ has a distribution centered on the population mean μ is what makes the sample mean a natural choice for approximating the population mean. However, we know that a random variable need not take on its mean value in any particular case, so there remains the possibility that in our particular sample the sample mean will turn out to be uncomfortably far from the population mean. To get a sense of how far off a sample mean can be we consider its dispersion, as summarized in the standard deviation of the sample mean.

The sample mean has standard deviation $\sigma/\sqrt{n}$ when the population has standard deviation σ. When the sample size $n = 100$, for instance, the sample mean has a standard deviation only one-tenth as large as the population standard deviation. Thus any given sample mean is likely to turn out to be closer to the population mean than is any individual case, since very low and very high values tend to cancel out when averages are taken. Figure 7.1 compares the distribution of a single case with the

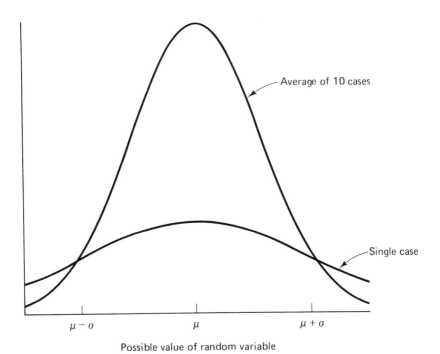

Possible value of random variable

Figure 7.1
Comparing the distribution of the value of a single case with that of the average of
10 cases

distribution of the average of ten cases. Note that the distribution of
the average is more concentrated around the value of the population
mean μ. To emphasize that the sample mean is used as an estimate of
the population mean, statisticians have come to call the standard deviation
of the sample mean the *standard error*.

Knowing that $\bar{x}$ is a Gaussian random variable with mean μ and
standard deviation (or standard error) $\sigma/\sqrt{n}$, we can write the likelihood
of the observed sample result as

$$\text{Likelihood } [\bar{x} \mid \mu, \sigma, n] = \frac{1}{\sqrt{2\pi}\, \sigma/\sqrt{n}} \exp\left[-\frac{1}{2}\left(\frac{\bar{x} - \mu}{\sigma/\sqrt{n}}\right)^2 \right],$$

emphasizing that the likelihood is conditional on the values of μ, σ, and
n. Now in any application we will know the sample size n, but μ and σ

pose problems. The population mean μ is the unknown we seek to estimate, but σ will generally be unknown as well. The population standard deviation

$$\sigma = \sqrt{\frac{\sum_{i=1}^{N} (x_i - \mu)^2}{N}}$$

cannot be computed since we do not have available all N values of the variable x. However, we can use as an approximation to σ a quantity of similar form known as the *sample standard deviation*:

$$S = \sqrt{\frac{\sum_{i=1}^{n} (x_i - \bar{x})^2}{n - 1}} .$$

We can compute S using only the observed sample values, and for large n the value of S will be a good estimate of σ. Hence for large samples the likelihood function will be well approximated by a Gaussian distribution centered on $\bar{x}$ with standard deviation approximately $S/\sqrt{n}$:

$$\text{Likelihood } [\bar{x} \,|\, \mu, S, n] \approx \frac{1}{\sqrt{2\pi}\, S/\sqrt{n}} \exp \left[-\frac{1}{2} \left(\frac{\bar{x} - \mu}{S/\sqrt{n}} \right)^2 \right].$$

Ideally, the likelihood function would be highly concentrated, meaning that only a narrow range of possible values of the unknown μ would be consistent with the data, and our uncertainty of estimate would be small. This will be the case whenever the standard error $\sigma/\sqrt{n}$ (as approximated by $S/\sqrt{n}$) is small: whenever σ is small or n is large or both. In other words, if the population is rather homogeneous (small σ) so that any one case is rather like any other, even a small sample is likely to yield a sample mean close to the mark. (In the extreme, if all cases are exactly alike, a sample of size $n = 1$ tells us everything). Even if the population is rather heterogeneous (large σ), we can minimize our uncertainty by taking a correspondingly larger sample (large n). You should be aware, though, that there are decreasing returns to scale that make it rather expensive to do away with the last vestiges of uncertainty. Because the standard error is inversely related to the square root of the sample size, it takes a quadrupling of the sample size n just to cut the standard error in half.

The Likelihood Function for Small Samples

We stressed in the last section that the Gaussian approximation to the likelihood function of the unknown population mean μ would really only be adequate for large samples (on the order of 50 or more cases). Very often in planning practice there will only be available a small sample, such as the sample of full-value tax rates in 11 towns in table 7.1. In such small samples it is no longer safe to act as if the sample standard deviation S is an excellent approximation to the unknown population standard deviation σ. We must pay the price for having to approximate σ, and that price is a likelihood function more diffuse, less definite than the Gaussian; this new likelihood function is known as *Student's t distribution with n — 1 degrees of freedom*. (The technical details of the notion of degrees of freedom are unimportant here: you can just think of degrees of freedom as a way to distinguish t distributions arising from samples of different sizes.)

Student's t distribution acquired its name through William Gossett, a statistician in the employ of the Guinness breweries around 1900, who used the pseudonym "Student" when reporting his pioneering work on sample means. Student knew that if simple random samples of size n were drawn from a Gaussian population with mean μ and standard deviation σ, then the sample mean $\bar{x}$ would itself be a Gaussian random variable with the same mean μ and a smaller standard deviation $\sigma/\sqrt{n}$. Hence the standardized version of the sample mean

$$Z = \frac{\bar{x} - \mu}{\sigma/\sqrt{n}}$$

would be Gaussian with mean 0 and standard deviation 1.0. However, since σ must be approximated by the sample standard deviation S, anyone working from a sample could not compute Z; rather the analyst would compute the corresponding quantity, which Student called t

$$t = \frac{\bar{x} - \mu}{S/\sqrt{n}}.$$

Student gave the exact formula for the distribution of t, which turns out to depend on the sample size n. As the sample size increases, S becomes a better approximation for σ, the quantity t grows closer to Z, and Student's t distribution becomes indistinguishable from the Gaussian, justifying our Gaussian approximation for large samples.

The actual form of the likelihood function is

$$\text{Likelihood } [\bar{x} \,|\, \mu, S, n] \; \alpha \left[1 + \frac{n(\bar{x} - \mu)^2}{(n-1)\,S^2} \right]^{-n/2}$$

which peaks at $\mu = \bar{x}$ and decreases symmetrically. Student's t distribution is flatter and wider than the Gaussian, honestly reflecting the additional fuzziness introduced into our estimate of μ by our uncertainty about the dispersion of the population distribution. Returning to the

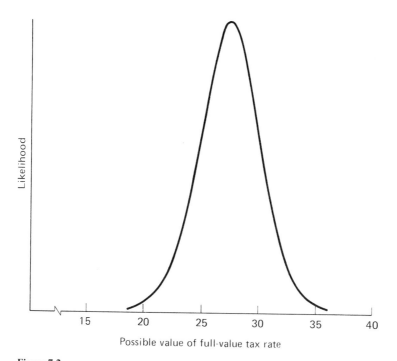

Figure 7.2
Estimation of full-value tax rate from a sample of 11 communities

sample of tax rates in table 7.1, we note that $n = 11$, $\bar{x} = \$27.42$, and $S = \$8.59$, so the likelihood for that example is

Likelihood $[\bar{x} = \$27.42 \,|\, \mu, S = \$8.59, n = 11]$
$\quad \alpha \,[1 + 0.0149(27.42 - \mu)]^{-5.5}$.

This function is shown in figure 7.2, from which we conclude that the statewide average full-value tax rate is unlikely to be less than \$20 or more than \$35. If we need more precision in our estimate we must sample more towns. Out best guess, of course, is $\bar{x} = \$27.42$, at least until we obtain more data.

A review of the steps involved in estimating a population mean is given in table 7.2.

Highest Density Regions for the Population Mean

Just as when estimating a probability, we may on occasion desire a less informative but more succinct summary of the uncertainty of estimate

Table 7.2
Steps in determining the posterior distribution of the population mean as estimated from a sample

1. Specify a prior distribution for population mean μ.

2. Draw a simple random sample of n cases from the population (which is assumed to be Gaussian and much larger than n).

3. Determine the value of the random variable of interest X in each of the n cases.

4. Compute the sample mean

$$\bar{X} = \frac{1}{n} \sum_{i=1}^{n} X_i.$$

5. Compute the sample standard deviation

$$S = \sqrt{\frac{1}{n-1} \sum_{i=1}^{n} (X_i - \bar{X})^2}.$$

6. Pick a possible value of the unknown population mean μ.

7. Compute the likelihood

$$\text{Likelihood } [\bar{X} \,|\, \mu, S, n] \; \alpha \left[1 + \frac{n(\bar{X} - \mu)}{(n-1)S^2} \right]^{-n/2}$$

8. Multiply the likelihood by the value of the prior distribution at the chosen value of μ to obtain a point on the posterior distribution.

9. Go to step 6.

than that provided by the full posterior distribution. In such cases we may cite a highest density region (*HDR*) for the unknown population mean. When we use a flat prior so that the posterior distribution takes the form of the likelihood, the tables of percentage points of Student's *t* distribution in appendix C help establish the upper and lower margins of the *HDR*.

For instance, consider again the sample of tax rates in $n = 11$ Massachusetts towns, where we found a sample mean $\bar{x} = \$27.42$. Suppose we want to establish a 95 percent *HDR* for the unknown population mean μ. We would determine the 97.5 percent point on Student's *t* distribution with $n - 1 = 10$ degrees of freedom, which happens to be $+2.228$. Since the *t* distribution is symmetric, the 2.5 percent point is -2.228, so 95 percent of the area under Student's *t* distribution with 10 degrees of freedom lies between -2.228 and $+2.228$,

$$\text{Prob}\,[-2.228 \le t_{df=10} \le +2.228] = 0.95.$$

Now we know that

$$t = \frac{\bar{x} - \mu}{S/\sqrt{n}},$$

so

$$
\begin{aligned}
0.95 &= \text{Prob}\,[-2.228 \le \frac{\bar{x} - \mu}{S/\sqrt{n}} \le +2.228] \\
&= \text{Prob}\,[-2.228\,S/\sqrt{n} \le \bar{x} - \mu \le +2.228\,S/\sqrt{n}] \\
&= \text{Prob}\,[-\bar{x} - 2.228\,S/\sqrt{n} \le -\mu \le -\bar{x} + 2.228\,S/\sqrt{n}] \\
&= \text{Prob}\,[\bar{x} + 2.228\,S/\sqrt{n} \ge \mu \ge \bar{x} - 2.228\,S/\sqrt{n}].
\end{aligned}
$$

Substituting the values of $\bar{x}$, S, and n for the tax rate example, we get

$$\text{Prob}\,[\$33.19 \ge \mu \ge \$21.65] = 0.95.$$

Thus the 95 percent *HDR* for the unknown μ ranges from $\$21.65$ to $\$33.19$. If we were willing to settle for a less conservative *HDR*, we could cite a correspondingly narrower range; for example, an 80 percent *HDR* (there is four times as much chance that μ will lie within the *HDR* as outside it) would be $\bar{x} \pm 1.37\,S/\sqrt{n} = \$27.42 \pm \$3.55$ or $\$23.87$ to $\$30.97$.

In general the limits of the *HDR* for the population mean are given by the expression

$$\bar{x} \pm C_{n-1} \times S/\sqrt{n},$$

where C_{n-1} is a constant which depends on the sample size and the degree of confidence we wish to attach to the proposition that the unknown value of μ lies within the HDR. For a sample of size n and a P percent HDR, C will be the $(100 + P)/200$ percentile of Student's t distribution with $n - 1$ degrees of freedom. (See appendix C.)

Choosing the Sample Size

The simple formula for the HDR is helpful when thinking about choosing the size of the sample to be drawn from the population. The key factor to consider is the heterogeneity of the population: if all cases are identical, a sample of size $n = 1$ will suffice; the more heterogeneous the population, the larger the sample needed to produce a narrow posterior distribution or HDR.

Suppose we wanted the width of the 95 percent HDR in the tax rate example to be no larger than $5.00. If we knew the population standard deviation σ, then the likelihood function would be Gaussian and the 95 percent HDR would be given by $\bar{x} \pm 1.96\,\sigma/\sqrt{n}$, since 1.96 is the 97.5 percent point of the standard Gaussian distribution. Accordingly the width of the 95 percent HDR would be twice $1.96\,\sigma/\sqrt{n}$, and we would determine the minimum acceptable sample size n from the inequality

$$2(1.96\sigma/\sqrt{n}) < \$5$$

or

$$n > \frac{2(1.96\sigma)^2}{5}.$$

However, in practice we will usually not know σ.

We might proceed in one of two ways. First we might use our substantive knowledge to estimate σ, erring on the high side to be safe, then use the inequality above. Alternatively, we might obtain a small exploratory sample and use the sample standard deviation S as an approximation to σ. In this case the likelihood has Student's t distribution, and we would chose a value of n such that

$$2(C_{n-1}\,S/\sqrt{n}) < \$5.$$

Since C_{n-1} will be about 2.0 for sample sizes $n > 30$ or so, we can make a quick estimate using

$2(2S/\sqrt{n}) < \$5$

or

$$n > \left(\frac{4S}{5}\right)^2.$$

Since the sample of $n = 11$ towns in table 7.1 produced the sample standard deviation $S = \$8.59$, we would estimate that a sample size

$$n > \left(\frac{4 \times 8.59}{5}\right)^2 = 47$$

would be about large enough to provide an estimate of sufficient precision.

In general, if we wish the width of the 95 percent HDR to be no wider than W, we should plan on a sample size of at least

$$n > \left(\frac{4S}{W}\right)^2,$$

where the value of S is calculated from an exploratory sample if possible or estimated subjectively if necessary. This rule of thumb should work well provided (1) the prior distribution of μ is flat and (2) the resultant sample size n is not very large compared to the total number of cases in the population. If either proviso (1) or (2) or both are not true, the estimated sample size will probably be larger than necessary, but even an overestimate is useful to have.

We have been discussing the issue of sample size as if the planner had control over the sampling process. It will often happen that the sample has already been drawn by someone else, perhaps for another purpose, and further sampling is impractical. In such situations the issue of choice of sample size is moot. You should focus your attention in such cases on whether the sample was actually a simple random sample. If not, then the methods of this chapter are not strictly valid and may be most inappropriate.

Sampling Experiments

We know now how to process a particular sample and convert the data to a posterior distribution or a highest density region for the population mean. It will be useful to cultivate a feel for the variations arising from sample to sample by playing with some synthetic data. In practice you will normally be dealing with a single sample and will have no chance to see what a second or third sample of the same size from the same population would look like, but taking many samples from the same population is an academic experience of great value.

Our data source is a table of standardized Gaussian random numbers. Recall that the standardized Gaussian distribution has mean zero and standard deviation unity. We will generate many small samples from this distribution to learn

1. what samples from a Gaussian population look like,
2. how much the sample means differ from the population mean of zero,
3. how often the *HDR*'s actually bracket the population mean,
4. how much improvement comes from taking larger samples.

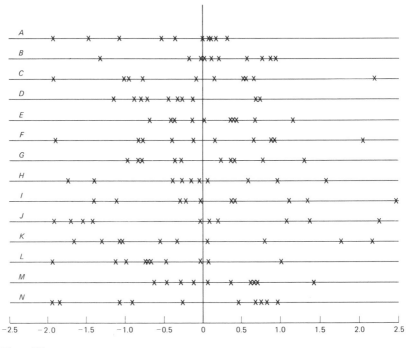

Figure 7.3
Plots of samples of 10 standard Gaussian random variables

Fourteen samples of size $n = 10$ are plotted in figure 7.3 (the data
are from Schmitt, p. 371). Recall that each is a sample from a population
whose mean is zero and whose standard deviation is unity. As you
examine each sample in turn, note that you never see a bell-shaped curve;
rather, you see scattered data points, each X signifying the value taken
on by the random variable for a case. Often the X's would hardly be
suspected of coming from a Gaussian population: sample A suggests
a population distribution bunched near zero and trailing off to the left,
while sample J looks like it might have come from a distribution with
three modes, near 0, -1.5, and $+1.5$. Clearly, it is very risky to make
inferences about a population distribution from a small sample (sample
H looks like the best behaved of the lot).

So what should you do in real life if you are presented, say, with
sample E of figure 7.3? You know that use of Student's t distribution
is based on the assumption that the population is Gaussian, but does
sample E justify the Gaussian assumption? In fact, no set of random
numbers is perfectly Gaussian—not even the artificial ones labeled as
Gaussian in the statistics books. Happily, inference based on Student's
t works rather well even when the population distribution is not quite
Gaussian: symmetry is important but exact normality is not (How do
we know this? By generating many samples from various non-Gaussian
distributions on the computer, then applying procedures based on
Student's t and comparing the results to those expected under ideal
conditions.) Most planners faced with sample E would go ahead and
use Student's t; I would advise its use but not without first eyeballing
the data. If an outrageous pattern occurs, such as one with all the values
clustered into two widely separated bunches or one that looked especially
asymmetrical, then the t distribution cannot be used. Rather other
methods would be tried that do not require the Gaussian assumption
(called "nonparametric methods," see Mosteller and Rourke's book),
or sometimes the data can be transformed by working, say, with logs
or square roots which might warp the data into a Gaussian pattern.
It would be helpful to remark when reporting the results that the data
appeared to be non-Gaussian. There are statistical tests (one is known
as the chi-square test) available to check the Gaussian assumption
formally. There is also a handy graphical test which we will describe.

Returning to our perusal of the samples in figure 7.3, let us check what kinds of errors we could make by asserting that the sample mean is close to the population mean (and that the sample standard deviation is close to the population standard deviation). Table 7.3 lists for each of the 14 samples the sample mean (which ideally would be 0), the sample standard deviation (which ideally would be 1.00), the 95 percent *HDR* (which ideally would bracket the point 0 in about 13 of the 14 samples) and the 50 percent *HDR* (which ideally would bracket the point 0 in about 7 of the 14 samples), computed assuming a flat prior.

Note the sample mean $\bar{x}$ never once took on the value of the population mean (0). In sample *L* it was rather far off the mark, while in sample *G* it was nearly right on. You can see that there is an element of luck involved. There are two ways to deal with this uncertainty inherent in the sampling process: first, take larger samples if you can afford to; second, express the results in terms of posterior distributions or highest density regions. By the way, note that just as the sample means buzzed around the right answer but never quite hit it, so too the sample standard

Table 7.3
Summary of results for samples of size 10 from a standard Gaussian distribution

Sample	Sample mean[a]	Sample standard deviation[b]	95 percent HDR		50 percent HDR	
			From	To	From	To
A	−0.46	0.77	−1.01	0.09	−0.63	−0.29
B	0.21	0.67	−0.27	0.69	0.06	0.34
C	−0.06	1.16	−0.89	0.77	−0.32	0.20
D	−0.32	0.63	−0.77	0.13	−0.46	−0.18
E	0.15	0.57	−0.26	0.56	0.02	0.28
F	0.07	1.12	−0.73	0.87	−0.18	0.32
G	−0.01	0.75	−0.55	0.92	−0.18	0.16
H	−0.08	0.99	−0.79	0.63	−0.30	0.14
I	0.18	1.10	−0.61	0.97	−0.06	0.42
J	−0.16	1.44	−1.19	0.87	−0.48	0.16
K	−0.11	1.31	−1.05	0.83	−0.40	0.18
L	−0.55	0.79	−1.11	0.01	−0.72	−0.38
M	0.24	0.64	−0.22	0.70	0.10	0.38
N	−0.23	1.13	−1.04	0.58	−0.48	0.02

[a] Population mean = 0
[b] Population standard deviation = 1.00.

deviation buzzed around the population value of 1.00 but never quite hit it either (although sample H certainly came close).

Now pay attention to the 95 percent HDR and the 50 percent HDR. (Check in tables of Student's t to confirm for yourself that the 50 percent HDR for samples of size $n = 10$ is computed from $\bar{x} \pm 0.70S/\sqrt{10}$.) If all is right in the world, we would expect the 95 percent HDR to include the population mean in 95 of every 100 samples. In our set of 14 samples we expect the 95 percent HDR to include the population mean $\mu = 0$ in something like $0.95 \times 14 = 13.3$ cases. In fact, all 14 of the 95 percent HDR's include $\mu = 0$, so we have good agreement between the theory and our data. An even better test is provided by the 50 percent HDR's, since we expect more variation in results: on average, 7 of 14 samples should have 50 percent HDR's that include the population mean. In

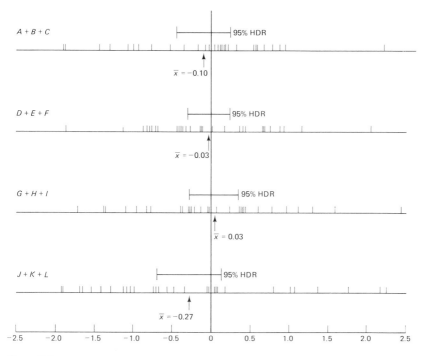

Figure 7.4
Plots of samples of 30 Gaussian random variables showing sample means and 95 percent HDR's for population means

fact, 8 of the 14 include $\mu = 0$; again we have good agreement. To make a final point about highest density regions, note in table 7.3 that for such small samples the 95 percent HDR's are rather wide (relative to the sample standard deviation): with small samples it is often difficult to pin down the value of the population mean.

Now let us see what is gained by taking larger samples. We can look at samples of size $n = 30$ by combining our original samples three at a time. These larger samples are plotted in figure 7.4. Note that they look much more Gaussian than do the samples of size $n = 10$: relatively symmetric, with most cases clustered near zero and a general thinning out as the values become more extreme. Figure 7.4 also shows the sample means and the 95 percent HDR's (for samples of size $n = 30$, 95 percent $HDR = \bar{x} \pm 2.045 \ S/\sqrt{30}$); the sample means are close to the nominal population mean $\mu = 0$, and the HDR's bracket $\mu = 0$ in all cases. The 95 percent HDR's for samples of size $n = 30$ are only about half as wide as those for the samples of size $n = 10$, but of course involve more time and money to obtain.

Testing the Gaussian Assumption

We outline now a simple graphical procedure for informally testing whether any given set of data points arises from a Gaussian distribution.

Table 7.4
Ordered data from sample H in firgure 7.4

Order i	Value X_i	$i/n + 1$
1	−1.91	1/11 = 0.09
2	−1.45	2/11 = 0.18
3	−1.07	3/11 = 0.27
4	−0.51	4/11 = 0.36
5	−0.35	5/11 = 0.45
6	0.00	6/11 = 0.55
7	0.09	7/11 = 0.64
8	0.11	8/11 = 0.73
9	0.18	9/11 = 0.82
10	0.31	10/11 = 0.91

The key is the use of so-called "probability paper" (costing about 3¢ per sheet) which will show the data along a straight line if they are Gaussian. The procedure is

1. Get a sheet of probability paper.
2. Get a sample of data, say, n cases.
3. Arrange the n data items in order from smallest to largest. Call the ith smallest x_i.
4. Plot the pairs $\left(x_i, \dfrac{i}{n+1} \right)$ on the probability paper.
5. If the points fall on a reasonably straight line, then the data are approximately Gaussian.

To work an example, begin with sample H from figure 7.3. The data, arranged in order, and the corresponding values of $i/(n+1)$ are listed in table 7.4. These data are plotted on probability paper in figure 7.5, as are the theoretical straight line and the pairs from a sample of size $n = 30$ (samples $G + H + I$ merged). Note how much more closely the larger sample plots to the theoretical straight line. Even in the case of the larger sample, though, the end points (the *tails*) tend to diverge from the straight line. Remember that the samples really are from a Gaussian population; even the small sample of $n = 10$ cases, should, in theory, be close to the straight line. By making such plots or performing other sampling experiments with synthetic data you can do much to develop your intuitions about the typical kinds of discrepancies that arise in practice between theoretical models and actual sample results. (For a formal method of testing the Gaussian assumption, see Mosteller and Rourke, chapters 8 through 10.)

Alternatives to the Sample Mean

Planners take sample averages so naturally and so often that they often fail to realize that there are many other ways to estimate the mean of a population. It happens that, if the population is Gaussian, the sample mean is the best estimate to compute from the sample data; since many populations are Gaussian or nearly so, the sample mean is therefore often a good estimate to use. However, if the population has a uniform distribution, it is preferable to use the average of the largest and smallest data values (the *midrange*). For distributions that are symmetric and

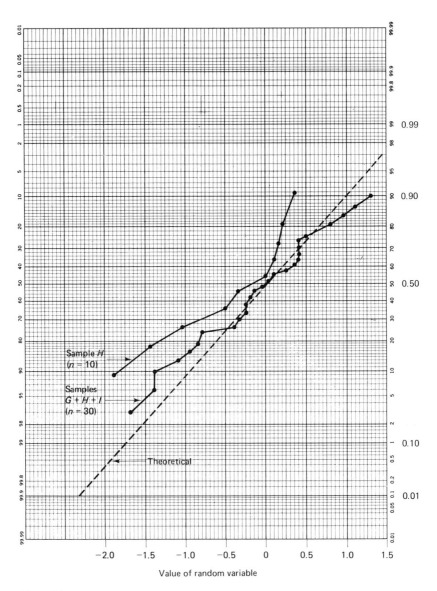

Figure 7.5
Plot of samples from the standard Gaussian on probability paper

have long tails, it is better to use the median of the sample to estimate the mean of the population. There are also techniques (known as *Windsorizing*) based on replacing the largest and smallest sample values with less extreme values to protect against deviant cases and outright errors in the data. The technique of *trimming* addresses the same worries by simply throwing out the largest and smallest values. You should be dimly aware at least of the existence of these alternatives; the sample mean is by no means the only possible estimate of the population mean.

Summary

One way to estimate the mean value of a random variable in a population is to draw a simple random sample of cases and compute the average of those cases. This sample average can be used as an estimate of the unknown population mean. Our concern in this chapter has been to explicitly display the uncertainty inherent in such an estimate. The full description of that uncertainty is contained in the posterior distribution of the unknown population mean, computed as the product of the prior distribution and the likelihood function. When the sample size is large (perhaps 50 or more) the likelihood function is well approximated by a Gaussian distribution with mean $\bar{x}$ (the sample mean) and standard deviation $S/\sqrt{n}$ (the standard error). When the sample size n is small, we have

$$\text{Likelihood } [\bar{x} \mid \mu, S, n] \propto \left[1 + \frac{n(\bar{x} - \mu)^2}{(n - 1)S^2} \right]^{-n/2}.$$

A concise way to express the uncertainty in the estimate is the 95 percent highest density region (HDR), which is simply the range of values that includes 95 percent of the area under the posterior distribution where every point is more likely to be the actual answer than any point outside. When the prior distribution is flat the 95 percent HDR is given by

$$\bar{x} \pm C_{n-1}S/\sqrt{n},$$

where C_{n-1} is the 97.5 percent point of Student's t distribution with $n - 1$ degrees of freedom. These formulas for small samples assume that the data consitute a simple random sample from a Gaussian population much larger than the sample. The Gaussian assumption may be tested informally by a graphical technique using probability paper. A good way to develop a sense of sampling variability is to draw repeated samples from a known distribution and observe the sample-to-sample differences.

References and Readings

Daniel, W. "Statistical Sampling and the Appraiser." *The Appraisal Journal* 43 (1975): 90–104.

Mosteller and Rourke. "Goodness of Fit for Continuous Distributions," section 10.4, "Confidence Limits for the Median and Other Quantiles of the Population," section 14.2, and "Ideas of Point Estimation for Means and Medians," section 15.1, pp. 185–186, 236–240, 248–250.

Rubin, D. "Formalizing Subjective Notions About the Effect of Nonrespondents in Sample Surveys." *Journal of the American Statistical Association* 72 (1977): 538–543.

Schmitt. "Nonnormality," chapter 7, pp. 203–230.

Williams, W. "How Bad Can "Good' Data Really Be?" *The American Statistician* 32 (1978): 61–65.

7.1

Data on per capita municipal expenditures in 20 Massachusetts communities are given in table 10.2. Graph the likelihood function of the mean per capita expenditure. What is the 95 percent *HDR* for the mean? About how many more communities would have to be sampled to reduce the width of the 95 percent *HDR* to $100?

7.2

Sometimes the sample median is superior to the sample mean as an estimator of the mean value of a symmetrical distribution (for a symmetrical distribution, the population mean and median are identical). The sample median will be more attractive when the random variable of interest has a distribution wider and flatter than the Gaussian. To conveniently simulate such a distribution, form samples of size $n = 3$ by drawing the first two sample values directly from a table of standard Gaussian random numbers and using as the third sample value a Gaussian random number multiplied by 5. This procedure should give more extreme sample values than would be produced by the Gaussian (the resultant distribution is known as a "contaminated" Gaussian since it is a mixture: 2/3 Gaussian with standard deviation 1.0 and 1/3 Gaussian with standard deviation 5). Form 10 of these samples of size $n = 3$. (a) Combine all 10 samples, and plot the 30 values on probability paper; does the plot suggest a Gaussian or non-Gaussian distribution? (b) For each of the 10 samples compute the sample mean and sample median and determine which is closer to the true value of the population mean. Using the Bayesian techniques of the previous chapter and a flat prior, use the results of the 10 comparisons to determine the posterior distribution of the probability that the sample median is a better estimate of the center of the distribution than the sample mean for the particular distribution simulated.

7.3

A planner seeks an estimate of the average rate at which residents of an apartment building for elderly citizens visit a branch library. A random sample of 10 residents yields the information reported in the table on the number of months the resident has lived in the building and the number of trips he has made to the library in that time.

a. The planner realizes that there are two ways to estimate the average rate of monthly visits per resident from the data. Calculate the two estimates.

b. In general, when will the two methods give the same result?

c. On the assumption that each resident's library visits are independent Poisson events with mean number $A = R \times T$, where R is the unknown rate and T is the observed time of residence in months, compare the likelihoods of the sample results using each of the two estimates of the rate. On this basis, which estimate seems preferable?

d. On what bases might you question the assumption that each resident's library visits are independent Poisson events?

e. What other approach might the planner have taken to estimate the mean rate?

Resident	Months in residence	Trips to library
1	1	2
2	48	10
3	6	3
4	15	1
5	4	0
6	13	5
7	25	9
8	10	1
9	3	1
10	7	0

III MAKING PREDICTIONS

8 Multivariate Analysis for Discrete Random Variables: Crosstabulations

We remarked in chapter 2 that planners usually desire conditional probability distributions as the bases for their predictions. When both the variable to be predicted and the attributes of the case that condition the prediction are discrete random variables, the crosstabulation or contingency table is the standard format for display of the conditional distributions.

In this chapter we will take up three basic issues in the analysis of crosstabulations. First, we will learn how to summarize the strength of association between two variables. Second, we will learn how to assess the view that an association observed in sample data may be nothing more than an accidental artifact of the sampling process. Third, we will learn how the association between a conditioning attribute and the variable being predicted can be clarified using additional attributes.

Measures of Strength of Association

We will generally be able to improve predictions about the value of a random variable if we can specify ahead of time the value of some associated attribute of the case; for instance, we might use average lot size or average education level or per pupil public education expenditures or some other attribute to predict the mean income in a community. However, when faced with a choice from among many attributes that might serve to condition out predictions, we would gain from having a summary measure of the predictive power of the attributes. Such a summary measure will provide an indication of the benefit to be had from investing effort in determining the value of an attribute and will make possible comparisons among attributes. We will discuss two approaches to measuring the predictive power of an attribute: proportional reduction in error and proportional reduction in uncertainty.

Proportional Reduction in Error

For a discrete random variable, our prediction will be a choice of the category taken on by the random variable of interest. The percentage of cases for which we predict the wrong category will measure our error rate. If we can use attribute information to condition our predictions so as to predict the right category more often, then that attribute is a useful

Table 8.1
Comparison of applicants' ratings by two professors

		Willemain rating			Row sum
		Admit	Maybe	Reject	
	Admit	2	2	0	4
Jones rating	Maybe	5	19	8	32
	Reject	0	1	8	9
	Column sum	7	22	16	45

predictor variable. The proportional reduction in error measures the predictive value of an attribute relative to unconditional predictions made without the attribute. In terms of a crosstabulation, if knowing which row of the table a case falls into improves our prediction of which column the case falls into, then the row variable is a useful predictor of the column variable.

Consider the crosstabulation in table 8.1, which displays some of the decisions made by myself and a colleague (whom I will call Professor Jones) during the process of admitting the Master of City Planning class of 1979 at MIT. A total of 45 applicants' folders were independently reviewed by both of us and classified into three groups: candidates clearly worthy of admission ("admits"), candidates who clearly should be rejected ("rejects"), and candidates who might merit admission upon further examination ("maybes"). Suppose the random variable of interest is my decision, and Professor Jones's decision is offered as an attribute that may help in predicting my decision.

Knowing only the overall distribution of my decisions, the best prediction of my judgment in any particular case would be "maybe," since "maybe" is the modal category among my decisions. Using such a prediction rule would lead to correct predictions in 22 of 45 cases and to incorrect predictions in $7 + 16 = 23$ cases, for an error rate of 51 percent.

Now suppose Professor Jones's judgment about an applicant were known to you before you made your prediction of my judgment in the same case. By how much would this extra information reduce your prediction error rate? To the extent that Professor Jones and I share a common set of admissions standards, knowing his decision should tell you a good deal about mine. If Professor Jones feels the applicant should definitely be rejected, reference to table 8.1 shows that the best prediction of my decision *conditional* on Jones's decision is "reject." Such a rule will lead to error in 1 of 9 cases, for an error rate of only 11 percent. Thus knowing that Professor Jones disapproves of an applicant is a powerful bit of information for predicting my response to the same applicant. If Professor Jones classifies the applicant as a "maybe," the best prediction of my decision is also "maybe." In this case the prediction rule leads to errors in $5 + 8 = 13$ cases, for an error rate of 41 percent, which is high but still less than the *unconditional* error rate of 51 percent when Jones's decision is unknown. Finally, when Jones considers the applicant a definite "admit," the best rules are to predict either "admit" or "maybe" as my judgment; both give an error rate of 50 percent. Thus the unconditional error rate is 51 percent, and the conditional error rate is 11, 41, or 50 percent depending on which category Professor Jones has chosen for the case in question. We can develop a single summary statistic for the conditional error rate by taking the average for all cases. Since Professor Jones classified 4 of the 45 applicants as "admits," 32 as "maybes," and 9 as "rejects," the average conditional error rate is

11 percent $\times$ (9/45) + 41 percent $\times$ (32/45) + 50 percent $\times$ (4/45)
    = 36 percent.

Finally, the *proportional reduction in error* is

(51 percent $-$ 36 percent)/51 percent = 0.30,

meaning 30 percent of the error rate in predicting my judgments has been eliminated by exploiting the conditioning information available in Professor Jones's ratings of the same applicants.

The proportional reduction in error statistic is known as λ (Greek lambda) and defined by the formula:

$$\lambda = \frac{\text{Prob [error]} - \sum_k \text{Prob [error} \mid \text{attribute} = k] \, \text{Prob [attribute} = k]}{\text{Prob [error]}},$$

where

Prob [error] = probability of error based only on the unconditional distribution of the random
variable of interest,

Prob [attribute = k] = probability that the conditioning attribute
takes the value k for any particular case,

Prob [error | attribute = k] = probability of error conditional on knowing that the attribute takes the value k.

There is an equivalent formula that will allow you to compute λ just by looking at a contingency table like that in table 8.1, when using the row variable to predict the column variable:

$$\lambda = 1 - \frac{\sum_{\text{rows}} (\text{off-mode counts in row})}{\sum_{\text{column totals}} (\text{off-mode counts in column totals})}.$$

For instance, in table 8.1 predicting my judgments from Jones's, we calculate

$$\lambda = 1 - \frac{(2 + 0) + (5 + 8) + (0 + 1)}{(7 + 16)} = 0.30.$$

Thus λ is an easy-to-compute measure with a straightforward and practical interpretation in terms of prediction error. It takes on values between 0 and 1.0, larger values signifying that the attribute chosen to condition the predictions is powerful at reducing prediction errors. (You probably have noticed that in applications like the admissions problem just discussed, λ can also serve as an *index of agreement*.)

Shown in table 8.2 are tables for which $\lambda = 0$ and $\lambda = 1.0$, using the row variable to predict the column variable. Note in the contingency table for which $\lambda = 0$ that the modal category is the same in every row, so knowing the value of the row variable never changes the prediction of the column variable; in this sense the association between the row and column variables is weak. On the other hand, in the contingency table for which $\lambda = 1.0$, knowledge of the value of the row variable not only changes the prediction of the column variable but makes for perfect predictions.

Lambda happens to be an asymmetrical measure in that the value computed using the row variable to predict the column variable need not

Table 8.2
Crosstabulations with proportional reduction in error λ equal to 0 and 1.0

	Predicted variable			
	8	10	9	
Predicting variable	0	5	1	$\lambda = 0$
	2	12	11	

	Predicted variable			
	5	0	0	
Predicting variable	0	0	8	$\lambda = 1.0$
	0	12	0	

equal the value computed the other way around. For instance, return to table 8.1, and this time consider using my judgments to predict those of Professor Jones, rather than the reverse. Using the quick formula, we compute

$$\lambda = 1 - \frac{(0 + 2) + (1 + 2) + (8 + 0)}{(9 + 4)} = 0.$$

Knowing my decision never changes the prediction that Professor Jones will classify the same applicant as a "maybe." However, there are shades of meaning in table 8.1 that may be important but are not reflected in the λ index: those I considered "accepts" were never rejected by Jones, and those I considered "rejects" were never accepted by Jones. A simple summary measure like λ (or $\bar{x}$ or any other) cannot carry all the information in a data set. Usually we learn more by also considering a second summary measure. The proportional reduction in uncertainty measures provide a useful complement.

Proportional Reduction in Uncertainty

In the proportional reduction in error view we gauged the predictive power of an attribute by the extent to which knowledge of the attribute reduced errors in prediction relative to the error rate without using the attribute. We can also take the view that the predictive power of an attribute depends on the extent to which its use reduces our uncertainty about which value the variable of interest will take on relative to our uncertainty without the attribute. The entropy of a distribution quantifies the notion of uncertainty.

Return again to table 8.1. Without knowledge of Professor Jones's rating of an applicant, you would have only the unconditional distribution of my ratings (the column totals) on which to base a prediction about my sense of an individual case. Since I did not react the same way to every applicant, there is some uncertainty as to which category any particular case will fall into. The entropy of the unconditional distribution of my ratings is

$$\text{Entropy [Willemain]} = -(\tfrac{7}{45} \log \tfrac{7}{45} + \tfrac{22}{45} \log \tfrac{22}{45} + \tfrac{16}{45} \log \tfrac{16}{45})$$
$$= 0.437.$$

This is a substantial degree of uncertainty, considering that the entropy would range from a minimum of 0 (if I had assigned every applicant to the same category) to a maximum of $\log 3 = 0.477$ (if I had assigned the applicants equally to the three categories).

Now consider each row of the contingency table. If you know that Professor Jones considers the case in question to be an "admit" (top row of table) then the uncertainty about my decision is

$$\text{Entropy [Willemain | Jones says "admit"]} = -(\tfrac{2}{4} \log \tfrac{2}{4} + \tfrac{2}{4} \log \tfrac{2}{4} + \tfrac{0}{4} \log \tfrac{0}{4})$$
$$= 0.301.$$

Likewise for the other rows

$$\text{Entropy [Willemain | Jones says "maybe"]} = -(\tfrac{5}{32} \log \tfrac{5}{32} + \tfrac{19}{32} \log \tfrac{19}{32} + \tfrac{8}{32} \log \tfrac{8}{32})$$
$$= 0.411,$$
$$\text{Entropy [Willemain | Jones says "reject"]} = -(\tfrac{0}{9} \log \tfrac{0}{9} + \tfrac{1}{9} \log \tfrac{1}{9} + \tfrac{8}{9} \log \tfrac{8}{9})$$
$$= 0.151.$$

In every case knowing Jones's judgment reduces the uncertainty about mine. Again we can average over all the cases to determine the average conditional entropy.

$$\text{Entropy [Willemain | Jones]} = 0.301(\tfrac{4}{45}) + 0.411(\tfrac{32}{45}) + 0.151(\tfrac{9}{45})$$
$$= 0.349.$$

Finally, the proportional reduction in uncertainty is given by *the uncertainty coefficient*

$$U = \frac{\text{Entropy [Willemain]} - \text{Entropy [Willemain | Jones]}}{\text{Entropy [Willemain]}}$$
$$= \frac{0.437 - 0.349}{0.437} = 0.20.$$

Hence knowledge of Professor Jones's decision reduces the uncertainty about my decision by 20 percent on average.

The general formula for the uncertainty coefficient is

$$U = \frac{\text{Entropy [distribution]} - \displaystyle\sum_{k} \text{Entropy [distribution | attribute} = k] \times \text{Prob [attibute} = k]}{\text{Entropy [distribution]}}$$

where

$$\text{Entropy [distribution]} = \text{entropy of the unconditional distribution,}$$
$$\text{Entropy [distribution | attribute} = k] = \text{entropy conditional on attribute taking the value } k \text{ for a particular case,}$$
$$\text{Prob [attribute} = k] = \text{probability that the conditioning attribute takes the value } k \text{ for a particular case.}$$

Like λ, the uncertainty coefficient ranges between 0 and 1.0, taking larger values when the attribute is a more powerful predictor.

The uncertainty coefficient is more difficult than λ to compute by hand (although both can be provided to you by the computer) and has a less operational interpretation. On the other hand, it is not so quick to declare "no association" between two variables. Whereas λ computed using my ratings to predict Professor Jones's failed to indicate the real association between our two sets of ratings (recall in that case $\lambda = 0$), the uncertainty coefficient does indicate an association:

$$U = \frac{\text{Entropy [Jones]} - \text{Entropy [Jones} \mid \text{Willemain]}}{\text{Entropy [Jones]}}$$

$$= \frac{0.339 - 0.251}{0.339}$$

$$= 0.26.$$

For work without a computer I prefer the convenience of λ, although I always look twice when I get the result $\lambda = 0$ and always request that U also be printed out when analyzing crosstabulations. In either case it proves useful to have a number ranging between 0 and 1.0 that summarizes the ability of an attribute to reduce uncertainty and improve predictions.

Assessing the Credibility of Apparent Associations in Samples

In the admissions example the data set consisted of all applications reviewed by both Professor Jones and myself during one academic year. The measures computed from the contingency table summarize how well the attribute information could reduce errors or uncertainty within that data set. Would the same prediction rules (if Jones says "reject," Willemain will say "reject") work as well during the next year's admission process? The only way to really answer this question is to try the rules again the next year. If the rules work well year after year, the continual confirmation will justify reliance on the rules. You may find an apparent association between two variables in one particular data set, but the best way to grow secure in its use for predictive purposes is to test the association in many times and places by replicating the analysis. You should always pay attention to whatever strong associations you may discover in a particular data set, but you should never take the results from just one data set to mean that you have discovered a powerful general rule. It may happen that the particular data set you analyze contains a purely fortuitous set of results that will never again be duplicated. Likewise, if you find little or no association in one data set, you need not conclude that a strong association could not exist elsewhere. There is always a temptation to make too much of a single analysis, perhaps because we often lack the energy or means to undertake another, but many a planner has been rudely surprised when an apparently strong association discovered in one data set disappeared in another context.

There are at least three reasons why associations discovered in one data set may disappear in another. First the world may change as the focus of analysis shifts to another place or another time or both. These changes often signal the presence of phenomena that should be added to the analysis as third variables. Second, there may have been a process of selection of cases for analysis that made the original data set unrepresentative. This can easily happen when samples are not drawn at random, so that only those cases easy to get at are included in the original analysis; such accidental samples are often severely biased. Third, it may be that even if random samples were drawn for the original analysis, plain bad luck may have produced an unrepresentative set of cases, giving the impression of a strong association when only a weak one in fact exists in the entire population. Of the three reasons why apparent associations may vanish upon further analysis, only the third receives much attention in most statistics courses, since it is the one that best lends itself to mathematical treatment. Nevertheless, you should be aware of the other two reasons and never forget the fundamental importance of replication of analyses as a source of confirmation.

The Statistical Significance of an Association in a Sample

In earlier chapters about estimating a count or a probability or a mean, we took pains to make explicit the uncertainty arising from the sampling process, relying on Bayes' rule to develop a posterior distribution displaying the relative credibility of all the possible estimates. You could imagine performing the same kind of analysis to estimate λ or U, beginning with a prior distribution between 0 and 1.0 (perhaps the beta distribution). Unfortunately, there is no known formula for the full likelihood function, so we cannot procede with a Bayesian analysis. Bayesian analyses of contingency tables are possible (see Lindley's paper) but beyond the scope of an introductory course. Instead we will settle for an incomplete, but still useful, analysis based on the notion of the statistical significance of an association discovered in a sample.

The thinking behind statistical significance testing is as follows: suppose you obtain a simple random sample of cases from a population, measure two discrete attributes, and form a contingency table comprised of counts of cases classified by the values of the two attributes. Computing some measure of the strength of association of the two attributes, you are intrigued to discover a value of either λ or U large enough to be of practical importance. However, you know that the value of λ or U for your contingency table depends on which particular cases were chosen from the population to make up your sample, and you wonder whether you might have accidently selected an unrepresentative set of cases that produce a sample value of λ or U much higher than that which you would have obtained had you analyzed the entire population. In other words, might you be overreacting, chasing a statistical will-o'-the-whisp which reflects nothing more than an accident of the sampling process?

Statistical significance testing is a way to partially address this fear of being made a fool of by a sampling accident. It provides some reassurance that the association you observed in the sample is indeed substantial by asking this question: How likely is it that such a strong association would arise just by accident when sampling from a population in which the two attributes were in reality completely independent? If there is a very small probability that an association as strong as that observed in the sample would arise from a population in which the two attributes were independent, then it becomes implausible to attribute the observed degree of association to a mere sampling accident. Correspondingly, belief in the existence of a nonzero association between the attributes in the population as a whole becomes more plausible.

On the other hand, if the association observed in the sample is of a magnitude that would arise often in the process of taking samples from a population having no real association between the attributes, then we can feel relatively confortable in dismissing the apparent association as an artifact of the sampling process.

Of course, there is more than one way to be made a fool of. While worrying about overreacting to a fluke result attributable to the sampling process, we might in fact underreact to a real association faithfully reproduced in the sample. Unfortunately, we generally have little chance of analyzing this obverse case mathematically, so there tends to be less attention paid to it by statisticians. Planners, however, may not be able to afford to miss that once-in-a-lifetime opportunity to exploit an association and should resist the general tendency in the statistical literature to fixate on the risk of overreacting to a sampling accident while ignoring the risk of underreacting to a real and possibly valuable association.

It is traditional to recast the decision about whether or not an observed association is real into a decision about whether or not to reject the hypothesis that no true association exists in the population as a whole, the observed association being attributable to sampling accident. This hypothesis is referred to as a *null hypothesis of independence*, since it argues that there is no association between the attributes. In formal significance testing, one must either reject or retain the null hypothesis after observing sample results that inevitably show some degree of association between the attributes. Now the null hypothesis may be either true or false and the decision about the null hypothesis either right or wrong, as shown in table 8.3. We make a correct decision either by retaining the null hypothesis when it is true or by rejecting it when it is false. Conversely, we can make an error either by underreacting and retaining the null hypothesis when it is false (sometimes called a "false positive" or "type II" error) or by overreacting and rejecting the null hypothesis when it is true (sometimes called a "false negative" or "type I" error). In general, we increase our vulnerability to one type of error when we attempt to reduce our vulnerability to the other, although we can reduce both risks by investing more time and money in a larger sample.

It is a common, and I believe unsatisfactory, practice to force the planner to retain or reject the null hypothesis by following some fixed rule determined before obtaining the sample results, such as, "If the degree of association observed in the sample is greater than such-and-

Table 8.3
Possible errors in testing the null hypothesis of independence

		True relationship in population	
		Association	No association
Inference from sample about null hypothesis	Reject	OK: correct inference	Error: overreaction to sampling accident
	Retain	Error: under-reaction to observed association	OK: correct inference

such, reject the null; otherwise retain the null." This stiff-legged approach often misleads the planner who is a statistical novice into believing that there is a magic value of some statistic computed from the sample that can unequivocally establish the truth or falsity of the null hypothesis. It is more faithful to the realities of the sampling process to recognize the ambiguities of sample results. Furthermore, the conventional approach tends to give the null hypothesis more attention than it deserves. A planner with substantive knowledge of the problem at hand will not usually bother to investigate attributes that really have no association; the constructive purpose of sampling and subsequent analysis is to discover and estimate useful relationships between variables, not to make a career of demolishing theories. The proper role for the null hypothesis is to serve as a foil to the planner's substantive theory of the association between two variables, to show how plausible it may be to explain an apparent association with the simpler theory that the association is merely an artifact of the sampling process.

It is this last, informal use which we shall adopt for significance testing. Our goal will be to estimate for any sample association its descriptive level of significance, defined as a conditional probability denoted by α (Greek letter alpha):

$$\alpha = \text{Prob} \left[\begin{array}{l} \text{sampling accident produces} \\ \text{association} \geq \text{observed} \\ \text{association} \end{array} \middle| \begin{array}{l} \text{null hypothesis of} \\ \text{independence is true} \end{array} \right].$$

To the extent that α is small, we can discount the possibility that the observed association is a sampling artifact. For reasons that have no special relevance to planning, it has become customary to consider results significant only if $\alpha < 0.05$, meaning that less than one time in 20 would simple random samples drawn from a population in which the two attributes are independent produce a degree of association as large as or larger than that observed in the particular sample actually drawn. Such small values of α strengthen the case for believing that the association observed in the sample is not an artifact, but they can never guarantee the point, and the planner who only takes an observed association seriously when the descriptive level of significance is 0.05 or less is at risk of occasionally overlooking the existence of a useful association.

I would recommend using the descriptive level of significance as an objective and important but not decisive factor in deciding whether to take an association between two variables seriously, and I would recommend ignoring the whole question of statistical significance unless the strength of association is great enough to warrant the attention in the first place.

The Chi-Square Test for Lack of Independence

The definition of the descriptive level of significance

$$\alpha = \mathrm{Prob} \left[\begin{array}{l} \text{sampling accident produces} \\ \text{association} \geq \text{observed} \\ \text{association} \end{array} \left| \begin{array}{l} \text{null hypothesis of} \\ \text{independence is true} \end{array} \right. \right]$$

requires that we have some measure of the strength of association between the variables in the contingency table. While λ or U could serve this purpose, we will introduce here the χ^2 (read "chi-square") statistic and show how to use it to estimate the value of α for a given table.

The chi-square test is one of the most commonly used significance tests. It is based on the fact that the existence of an association between two variables means they are not independent. The stronger the association, the greater the deviation from independence. Recall from chapter 2 that, when two attributes are independent, we can fill in the counts in each cell of the contingency table using only the row and column marginals:

$$\text{Expected cell count} = \frac{(\text{Row sum}) \times (\text{Column sum})}{(\text{Total count in table})}.$$

Thus we can readily compute for any contingency table the counts expected in each cell of the table under the null hypothesis of independence. When we observe counts that are very different from the expected counts, these discrepancies cast doubt on the validity of the null hypothesis. For the cell in the ith row and the jth column of the table, let the observed count be O_{ij} and the count expected under the null hypothesis be E_{ij}. The quantity $(O_{ij} - E_{ij})^2/E_{ij}$ is one of many possible measures of the discrepancy between the observed count in the cell and the count expected if the two attributes are independent. This particular measure is useful because it leads to a distribution of the overall discrepancy in the table that can be related to the Gaussian, but rather than establish the mathematical virtues of its form, we can note three features that make it intuitively attractive as an index of discrepancy. First, the squared difference in the numerator gives extra emphasis to extremely large discrepancies. Second, the squaring guarantees we can later add the terms from all the cells without fear of cancellations arising from occasional negative discrepancies masking the differences between expectation and observation. Third, the denominator term gives a sense of scale to the index, downplaying the importance of discrepancies that represent only small percentage deviations from what is expected under the null hypothesis.

Once we have recorded the observed cell counts O_{ij}, computed the expected cell counts E_{ij}, and computed for each cell its index of discrepancy $(O_{ij} - E_{ij})^2/E_{ij}$, we can reasonably form an overall index of discrepancy for the whole table simply by adding together all the discrepancies in the individual cells. This total summarizes in a single number the departure observed in the contingency table. We will call it χ^2_{obs} (Greek letter chi); for a table with R rows and C columns

$$\chi^2_{obs} = \sum_{i=1}^{R} \sum_{j=1}^{C} (O_{ij} - E_{ij})^2/E_{ij}.$$

We have reduced the full pattern of discrepancies in the table to a single number which will be small when the two attributes exhibit independence and large when they are strongly associated. (Note that χ^2_{obs} might therefore play the same role as λ or U in measuring strength of association, although it is less attractive than the other two measures in certain key respects.) A large value of χ^2_{obs} casts doubt on the null hypothesis of independence.

A key step in understanding the chi-square test or any other significance test is to realize that because of the sampling process there will inevitably be some degree of association between the attributes in the sample even if the null hypothesis is true for the population. Hence it will almost never happen that the observed counts O_{ij} exactly equal the expected counts E_{ij} in every cell of the table, and therefore it will almost always be true that $\chi^2_{obs} > 0$. Accordingly, the fact that χ^2_{obs} is nonzero does not by itself reflect doubt on the null hypothesis of independence, since, even when the null hypothesis is true, accidents of sampling will cause some departures from independence to appear in the contingency table.

On the other hand, such sampling accidents are unlikely to generate extreme discrepancies across the table, so at some point the argument that very large values of χ^2_{obs} can be dismissed as mere sampling artifacts begins to wear thin.

The most important piece of work in constructing a significance test is to derive an understanding of these variations in the measure of discrepancy from sample to sample. As usual when dealing with random variables, we summarize our knowledge in terms of a probability distribution. The statistic computed from the sample to test the null hypothesis is called the *test statistic*, and its distribution when the null hypothesis is true is called the *sampling distribution of the test statistic*. The value α of the descriptive level of significance is computed from the right-hand tail of the sampling distribution of the test statistic.

In the case of testing the null hypothesis of independence between two attributes in a contingency table, the test statistic is χ^2_{obs}. Since χ^2_{obs} is a function only of the observed cell counts O_{ij} (recall that the expected counts E_{ij} are themselves computed from the O_{ij}), and since the O_{ij} can take on only the discrete values 0, 1, 2, 3, 4, and so on, it follows that the exact sampling distribution of χ^2_{obs} will be a discrete probability distribution. However, only for small tables (2 rows and 2 columns and small cell counts) is it at all practical to work with the exact sampling distribution; in this case the test is called "Fisher's exact test" and uses the hypergeometric distribution (see problem 5.3). It is customary instead to use an approximation to the exact, discrete sampling distribution of χ^2_{obs} which is based on a continuous distribution called the *chi-square distribution with* $(R - 1) \times (C - 1)$ *degrees of freedom*, where R is the number of rows and C the number of columns in the table. Thus, for instance, in a table with 3 rows and 3 columns such as table 8.1, the test statistic has a distribution approximated by the chi-square distribution with $(3 - 1) \times (3 - 1) = 4$ degrees of freedom.

Such a distribution is shown in figure 8.1, which illustrates graphically that the descriptive level of significance α is the probability that the test statistic will equal or exceed the observed value when the null hypothesis is true. From figure 8.1 we learn that when dealing with tables with 3 rows and 3 columns (called "3 × 3" or "three by three" tables), sampling accidents will very often produce overall discrepancies in the approximate range 1.0 to 3.0, and discrepancies as large as 6.0 or 7.0 are not uncommon.

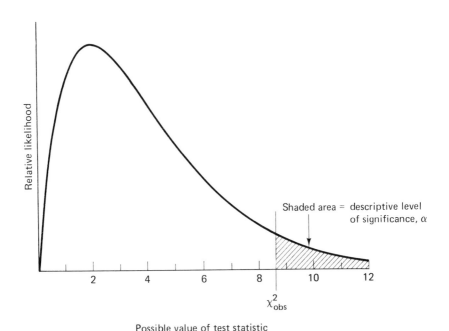

Figure 8.1
The chi-square distribution with 4 degrees of freedom, showing the descriptive level of significance

Therefore, only values of χ^2_{obs} of about 8.0 or more are hard to explain as sampling artifacts when working with 3×3 tables. Since larger tables have a greater number of cells, small discrepancies in each cell can add up to a larger but still innocuous discrepancy χ^2_{obs} for the table as a whole; accordingly, the standards for reacting to χ^2_{obs} shift with the size of the table (with the number of degrees of freedom). For a 5×5 table (degrees of freedom $= 16$), values of χ^2_{obs} as large as about 20.0 are rather unremarkable.

Most statistics books contain tables of the percentiles of chi-square distributions that make it possible to look up the descriptive level of significance of a value of χ^2_{obs}. Usually, however, χ^2_{obs} falls between two percentiles and must be interpolated within the table, and sometimes the tables are not handy, so we present here an approximation created by David Hoaglin that works well at approximating the right-hand tail of a chi-square distribution. For a table with $df = (R - 1) \times (C - 1)$ degrees of freedom and a total discrepancy χ^2_{obs}

$$\alpha \approx 10^{-\left(\frac{\sqrt{\chi^2_{obs}} - \sqrt{df + 7/6}}{2}\right)^2}.$$

Note that the above formula is an approximation to an approximation. Hoaglin's approximation to the tail of a chi-square distribution works especially well for tables with 4 or more degrees of freedom. The chi-square distribution as an approximation to the exact sampling distribution works best for tables having at least a few counts expected in every cell—the greater the number of counts, the better the approximation.

Example: Predicting Mortgage Lending from Housing Stock

It is time to work through an example in which measures of strength of association and tests of significance are calculated. The data were developed by Tee Taggart for the Massachusetts Banking Commission's study of alleged redlining (geographical discrimination in lending) by banks. The percentage of home sales financed by bank mortgages in 1975 to 1976 was recorded for 191 small areas (census tracts and zip codes) in Boston and environs, as were the proportions of total housing stock in 1970 comprised of older two- to four-family residences. Both the mortgage-to-sales and housing stock percentages were grouped into 3 categories and

Table 8.4

Association between bank lending and housing stock in 191 small areas

		Percentage of home sales financed by bank mortgages			
		Low (5–60)	Moderate (61–80)	High (81–92)	Row sum
Percentage of housing stock comprised of older 2 to 4 family homes	Low (<20)	4	25	41	70
	Moderate (20–50)	20	25	22	67
	High (>50)	35	12	7	54
	Column sum	59	62	70	191

Source: Harriett Tee Taggart, "Home Mortgage Lending Patterns in Metropolitan Boston," Commonwealth of Massachusetts Publication 10252-45-300-2-78-CR, December 1977.

presented for display in the 3×3 contingency table reproduced in table 8.4. (Incidentally, it is not generally recommended to treat metric data as if they were nominal, but the contingency table was chosen by Taggart as a convenient way to display the data for preliminary analysis. Her full analysis used the techniques of the next two chapters.)

Examination of the table shows a clear pattern: the composition of the housing stock is a good predictor of mortgage-lending activity in the areas, with more small multifamily units signalling less bank involvement in the financing of home purchases. We can objectively document and quantify this reading of the table by computing a measure of the strength of association between the two attributes. Then we can test the plausibility of the theory that an association of that magnitude could be merely a sampling artifact by performing a test of the statistical significance of the observed departure from independence.

In the absence of knowledge of the housing stock, our best guess for category of lending activity is high, since 70 of the 191 areas fall into that category; the error rate for this rule is $(191 - 70)/191$ or 63 percent. When we know nothing about the mix of housing types in the area, the entropy of the distribution of lending activity is very large:

$$\text{Entropy} = -(\tfrac{59}{191} \log \tfrac{59}{191} + \tfrac{62}{191} \log \tfrac{62}{191} + \tfrac{70}{191} \log \tfrac{70}{191}) = 0.476$$

compared to a possible maximum of $\log 3 = 0.477$. Our uncertainty can be reduced, however, and our predictions improved, if we use the con-

Table 8.5
Utility of housing stock information for predicting bank lending

Basis for conditional prediction: percentage of older 2 to 4 family homes in area	Conditional prediction: category of mortgage lending activity	Error rate for conditional prediction (percent)	Entropy of conditional distribution of lending categories
(Unknown)	(High)	(63)	(0.476)
Low	High	41	0.367
Moderate	Moderate	63	0.475
High	Low	35	0.382
		Mean = 47	Mean = 0.409
		$\lambda = 0.25$	$U = 0.14$

ditioning information available in knowledge of the housing stock, as shown in table 8.5. Using the conditioning information, we reduce our uncertainty by 14 percent ($U = 0.14$) and our error rate by 25 percent ($\lambda = 0.25$). While these improvements are not trivial, neither are they overwhelming, and it is appropriate to test the statistical significance of the association between the housing stock and mortgage-lending attributes.

Table 8.6 contrasts the observed counts with those expected in each cell under the null hypothesis that the attributes are independent. Note the large discrepancies in the four corner cells of the table. In areas with many older multifamily homes there is much less mortgage activity than would be expected under the hypothesis of independence, and much more activity than expected in areas with few such homes. The individual cells' contributions to the overall test statistic $\chi^2{}_{obs} = 54.66$ are shown in table 8.7. Hoaglin's approximation indicates that there isn't a snowball's chance in hell that an overall discrepancy as large as that observed could have arisen by accident of sampling, so we have great confidence that the association really exists and is not just an artifact.

The next steps in the analysis would probably search for extenuating circumstances, since the presumption after analyzing table 8.4 is that the practice advocated in most appraisal textbooks of avoiding investment in older multifamily dwellings is at work in the Boston area, creating a systematic pattern of discriminatory lending by banks. The extenuating circumstances would be defined in terms of other attributes of the areas under study; for instance, it would be worth checking whether areas with many older multifamily units also happened to be areas with high rates of foreclosure. We will take up soon the subject of basing predictions on more than one conditioning attribute.

A Word about Sample Size and Statistical Significance

The chi-square and all other significance tests give results that depend strongly on the number of cases analyzed. For instance, keeping the very same distribution of counts but doubling the total number will double the value of $\chi^2{}_{obs}$. This feature of significance tests can be useful if properly handled. A relatively small, but perhaps substantively important, association that might not be distinguishable from sampling fluctuations

Table 8.6
Comparison of observed counts with counts expected under null hypothesis of independence of housing stock and mortgage lending activity[a]

		Percentage of home sales financed by bank mortgages			
		Low	Moderate	High	Row sum
Percentage of housing stock comprised of older 2 to 4 family homes	Low	4 (21.62)	25 (22.72)	41 (25.66)	70
	Moderate	20 (20.70)	25 (21.75)	22 (24.55)	67
	High	35 (16.68)	12 (17.53)	7 (19.79)	54
	Column sum	59	62	70	191

[a] Expected counts are in parentheses. For example, expected count of areas with high percentage of older 2 to 4 family homes and low percentage of home sales financed by bank mortgages is

$$\frac{\text{Row sum} \times \text{Column sum}}{\text{Total count}} = \frac{54 \times 59}{191} = 16.68.$$

Table 8.7
Contributions of each cell to overall test statistic χ^2_{obs}[a]

		Percentage of home sales financed by bank mortgages		
		Low	Moderate	High
Percentage of housing stock comprised of older 2 to 4 family homes	Low	14.36	0.23	9.17
	Moderate	0.02	0.49	0.26
	High	20.12	1.74	8.27

[a] For example, for the middle cell $0.49 = (25 - 21.75)^2/21.75$; the total of all cell contributions is $\chi^2_{obs} = 54.66$; degress of freedom are $(3 - 1) \times (3 - 1) = 4$; the descriptive level of significance is

Prob $[\chi_{obs} \geq 54.66$ in table with d f $= 4 |$ null is true]

$$\approx 10^{-\left(\frac{\sqrt{54.66} - \sqrt{4 + 7/6}}{2}\right)^2}$$

$$= 2 \times 10^{-11}.$$

in a small sample will prove to be statistically significant in a sample of sufficient size. This is as it should be: we should not expect to be able to make fine distinctions with little empirical support, but a greater investment in data should pay dividends in terms of sensitivity to weak phenomena. Unfortunately there can be a disadvantage to the excruciating sensitivity that comes with very large sample sizes: any piddling association of almost no predictive power will turn out to be statistically significant even if it is of no practical importance. Especially when sample sizes are large, measures of the strength of association like λ or U will help you keep your reality orientation by identifying associations that are of no practical consequence even if they are statistically significant.

Making Predictions from Two or More Attributes

To this point we have concentrated on the problem of using a single attribute to predict another. We are naturally interested in bringing more information to bear by conditioning our predictions on more than one attribute. For instance, it might be useful to consider not only an area's mix of housing types but also its record of mortgage foreclosures when predicting banks' lending activity in that area. Most serious work will involve such *multivariate* analysis, which will often do much to further our understanding of interesting phenomena and improve our predictive power. We will illustrate the value of taking the time to include more attributes in the analysis, but we will also discuss in a general way the circumstances under which it is not worthwhile to add one more attribute.

First, to see how important it may be to include that additional attribute, consider the hypothetical data in table 8.8 relating rate of national economic development to degree of centralization in planning. The sample of 40 nations includes 22 developing and 18 developed nations. When the developed nations are analyzed separately, we see a negative association between centralized planning and rate of growth: less centralized developed nations tend to have more rapid growth rates. On the other hand, when the developing nations are analyzed separately, we see the opposite pattern: less centralized developing nations tend to have slower rates of growth. In each case there is an apparent association ($\alpha = 0.05$ for the developing nations and $\alpha = 0.01$ for the developed nations by Fisher's exact test) but of a very different type. However, if we merge

Table 8.8
A hypothetical illustration of the value of adding another attribute

Developed nations

		Rate of growth		
		Slow	Rapid	
Degree of	Less	4	8	(Negative
centralized				association)
planning	More	6	0	

Developing nations

		Rate of growth		
		Slow	Rapid	
Degree of	Less	6	2	(Positive
centralized				association)
planning	More	4	10	

Both groups combined

		Slow	Rapid	
Degree of	Less	10	10	
centralized				(No association)
planning	More	10	10	

all 40 cases into a bivariate analysis we see absolutely no link between centralization and growth. This hypothetical example was concocted to clearly illustrate how including an additional attribute (here, level of development) can enrich our understanding; without disaggregating the cases we would have concluded that there is no association between centralization and growth, when the data actually contain the message that there are two kinds of association.

If the number of developing countries had been doubled, the combined table would have displayed a positive association between centralization and growth; if the number of developed nations had been doubled, there would have been a negative association in the combined table. When there are distinct subgroups of cases (here developed and developing nations), the relative proportions of the subgroups in the sample can have powerful and misleading impacts on a simple bivariate analysis. If we have a substantive reason for distinguishing the subgroups, it is generally best to analyze them separately by adding another attribute that creates new tables.

There are two exceptions to this general rule, however. One arises when the sample size is such that further subdivision of the data would result in tables largely composed of empty cells. This would likely be the case for the data set shown in table 8.8, for which the counts are already uncomfortably low. If theory tells us that another attribute is crucial, then we must simply gather more data before we can safely push on with the analysis.

If certain values of an attribute are rare in the population, but it is important that we have several cases of that type for analysis, then we may need to augment the process of simple random sampling by *oversampling* for the rare cases. For instance, having drawn an initial simple random sample, we may discover that we have too few of those rare but important cases. We might then proceed with a further random search for just those cases, ignoring all others (remember there is a cost not just to selecting a case but also to obtaining data about that case, especially if there is an interview involved) and stopping when we have gathered enough to proceed reasonably with the analysis (which we hope will be before we run out of time, money, or patience). Such oversampling will result in a unrepresentative mixture of cases, but that matters little: our concern is with associations, not distributions (besides, the original simple random sample provides an estimate of the distribution of types of cases in the population if we need one).

The second exception to the general rule that it is useful to add one more attribute to the analysis arises when the new attribute is itself highly associated with attributes already in the analysis. This problem is generally referred to as *multicollinearity* and essentially means that little new information is provided by the additional attribute. Let us again examine hypothetical data to see the nature of the problem clearly.

The continue the redlining example, suppose a community group alleges that banks are engaged in systematic discrimination in mortgage lending against minority neighborhoods. The banking association, however, maintains that lending decisions are made solely on the basis of impartial economic analyses. A random sample of 20 neighborhoods provides data for an investigation. Each neighborhood is classified as minority or nonminority, its economic condition is rated as "good" or "bad," and

Table 8.9
A hypothetical illustration of the problem of multicollinearity

A. Neighborhood attributes showing extreme multicollinearity

		Economic conditions	
		Good	Bad
Racial composition	Minority	0	6
	Nonminority	14	0

B. Evidence supporting community group's allegation of racial discrimination

		Lending activity	
		Low	High
Racial composition	Minority	5	1
	Nonminority	4	10

C. Evidence offered by banks in their defense

		Lending activity	
		Low	High
Economic conditions	Good	4	10
	Bad	5	1

D. Attempt to isolate the effects of racial and economic factors

Good economic conditions

		Lending activity	
		Low	High
Racial composition	Minority	0	0
	Nonminority	4	10

Bad economic conditions

		Lending activity	
		Low	High
Racial composition	Minority	5	1
	Nonminority	0	0

the degree of bank mortgage activity in the neighborhood is classified as high or low. The data and analyses are shown in table 8.9.

The table in part A of table 8.9 shows immediately that the sample cases exhibit strong multicollinearity: there is a perfect association ($\lambda = 1.0$, $U = 1.0$, $\alpha = 0.0003$) between the racial and economic attributes. Every minority neighborhood in the sample has bad economic conditions, while every nonminority neighborhood in the sample has good economic conditions. While such an extreme degree of multicollinearity is not often found in a real sample, there is inevitably some degree of multicollinearity in any sample, so the problems evident in table 8.9 will be present to some extent in all the analyses you will undertake.

The community group will present the contingency table shown in part B of the table 8.9, arguing that the distinct ability to predict lending behavior from racial characteristics ($\lambda = 0.44$, $\alpha = 0.04$) is evidence of illegal discrimination. The banks will counter with the table shown in part C, arguing that it establishes a pattern of legal and prudent discrimination based on good business practice.

At this point it would be appropriate to conduct a multivariate analysis that examines the association between race and lending activity separately within each class of economic conditions. This step is often referred to as "controlling for a third variable," which in this case is the economic condition of the neighborhood. Unfortunately, such a move would produce the tables shown in part D. In both cases it is impossible to compare minority and nonminority neighborhoods because this particular sample contains no minority neighborhoods with good conditions and no nonminority neighborhoods with bad. The analysis can go no further until some neighborhoods are found to fill in the zeros in the table in part A. This attempt to add a theoretically important attribute to the analysis has failed because of multicollinearity; in the sample the data on economic conditions do nothing to modify the predictions based solely on racial makeup and therefore contain no useful information.

In an actual case of this sort, a more elaborate and careful analysis would clearly be in order. But even the best analysis can do no more than extract whatever information is encapsulated in the data, and if the data are anywhere near as redundant as those in the hypothetical example there can be little hope that an expensive analysis can save the situation. This is not to say, though, that we should limit ourselves to simple bivariate analyses. Usually we will do much better by bringing more attributes into the analysis—we should just not come to expect that our uncertainties will always be resolved by this strategy, or that good analysis can always compensate for bad data.

Using Associations to Search for Causality

To this point we have viewed contingency tables solely as a vehicle for improving predictions, although the previous example has eased us into a consideration of causality. In practice we are often interested in studying associations among attributes as a way of building theories about causal relations. Causal relations are the Holy Grail of analysis, since they promise the planner salvation through knowledge of the levers that directly control phenomena of interest. Although even with the best of luck it is never logically possible to establish causality beyond the shadow of a doubt, the pilgrimage in search of causality often leads to greater (if not perfect) knowledge of causal relationships, helping reasonable people make well-considered interventions.

It is important first to understand the distinction between observational studies and experiments. All of the examples used in this chapter arise from observational studies, be they real or hypothetical. An *observational study* records the world as it is normally, whereas an *experiment* records the world as it reacts to a change in the normal order. Therefore associations discovered in contingency tables arising from observational studies must be interpreted as follows: "If I *happen upon* a case for which attribute A has the value a, then I can expect to find that the case also has value b for attribute B." For instance, if an area happens to have a high proportion of older two- to four-family houses, it will likely be an area with little bank mortgage activity. This is essentially a *predictive* statement which in itself does not speak to the question of causality. The existence of the association does not prove that the nature of the housing stock causes the low level of bank lending for home purchase, although it is perhaps natural to begin suspecting so. In particular, it does not follow that changing the housing stock, perhaps by bulldozing all the small multifamily homes and replacing them with either open space or condominia, will necessarily change the level of banks' mortgage activity in the area. My friend and colleague Professor Thomas Nutt-Powell likes to recall that in his days as a community organizer in Baltimore, he always targeted his efforts to boost attendance at community meetings toward those people with house plants prominently displayed in their windows. Such people were usually receptive to the appeals of community organizers. This interesting association between the display of plants and attendance at meetings is a fine example of an association useful for predictive purposes but not for explaining causality: few would argue that the plants drove their owners to the meetings. When seeking causality, we refer to such an association as a *spurious correlation* (see chapter 9 for a discussion of the correlation coefficient); both keeping plants and attending community meetings probably derived from a certain 1960s ethos of plant owners at the time.

In contrast to observational studies, experiments involve recording the response to a change in the normal order. Without actually intervening in a system, it is never quite possible to make a solid case for a causal link between two attributes; even after conducting a planned change, it may be difficult to make the case, although an experiment is usually more enlightening (if sometimes more dangerous) than an observational study. We will take up the important subject of the design and analysis of experiments in chapter 11. Mosteller and Tukey make the point that a strong argument for causality usually requires (1) experimental evidence of *response* to change, (2) confirmation of that evidence by *replication* and (3) elucidation of a plausible causal *mechanism*. Unfortunately it is often difficult or impossible for the planner to perform experiments, so astute analysis of observational studies combined with solid professional judgment may be the best that can be hoped for.

To illustrate how multivariate analyses can help expose causal relationships not properly perceived in bivariate analysis, we consider three phenomena: spurious correlation, causal intermediaries; and multivariate causation. Professor Nutt-Powell's example of spurious correlation is illustrated with hypothetical data in table 8.10. A simple-minded causal interpretation of the bivariate analysis would conclude that the flowers in the windows cause attendance at the community meetings, but controlling for people's community spirit (somehow measured) shows that within each group those without flowers have the same probability of attending meetings as those with flowers. Thus a more plausible causal hypothesis is that both meeting attendance and flower cultivation arise

Table 8.10
A hypothetical illustration of the exposure of a spurious correlation

Bivariate analysis

		Attendance at meetings	
		Yes	No
Flowers in windows	Yes	90 (60%)	60 (40%)
	No	70 (35%)	130 (65%)

Multivariate analysis controlling for community spirit

Little community spirit
Attendance at meetings

		Yes	No
Flowers in window	Yes	10 (20%)	40 (80%)
	No	30 (20%)	120 (80%)

Much community spirit
Attendance at meetings

		Yes	No
Flowers in window	Yes	80 (80%)	20 (20%)
	No	40 (80%)	10 (20%)

Causal relationships

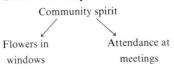

Community spirit

Flowers in windows   Attendance at meetings

from a common source in the consciousness of the people, as indicated by the directions of the causal arrows in table 8.10. An even more sophisticated theory would allow for *reciprocal causation* between community spirit and attendance at meetings, wherein each reinforces the other. In any event, controlling for community spirit demonstrated that the presence of flowers in windows did not impact on attendance. Examples in practice in which spurious correlation is an issue are numerous, although seldom as whimsical as Nutt-Powell's. For instance, in the redlining example of table 8.9 the community group might hold that the association between poor economic conditions and banks' inactivity arises because both poverty and lack of bank involvement derive from the racial makeup of the areas.

A second instance in which multivariate analysis clarifies causal relationships arises when two attributes are related through a third that operates as a *causal intermediary*. Infant mortality is known to be higher among illegitimate children, but does the legal status of illegitimacy directly cause the biological circumstance of greater risk? A more plausible theory holds that an important determinant of infant mortality is early prenatal care, and that a typical reaction to an illegitimate pregnancy is to delay the start of care, thereby increasing the risk of complications. Such a theory would be supported by evidence such as the hypothetical

Table 8.11
A hypothetical illustration of the exposure of a causal intermediary

Bivariate analysis

| | | Health outcome | |
		Live	Die
Legal status	Illegitimate	60 (92%)	5 (8%)
	Legitimate	105 (96%)	4 (4%)

Multivatiate analysis controlling for timing of prenatal care

Early prenatal care

| | | Health Outcome | |
		Live	Die
Legal status	Illegitimate	50 (98%)	1 (2%)
	Legitimate	100 (98%)	2 (2%)

Late prenatal care

| | | Health Outcome | |
		Live	Die
Legal status	Illegitimate	10 (71%)	4 (29%)
	Legitimate	5 (71%)	2 (29%)

Causal mechanism

data in table 8.11, in which the bivariate analysis shows a higher mortality rate among illegitimate babies, but the analyses controlling for timing of prenatal care show no distinctions between legitimate and illegitimate infant mortality rates. For both types of babies the risk is greater if care is delayed, and it is the more frequent delay that gives rise to the higher mortality rate for illegitimate children.

A third instance in which multivariate analysis helps expose causal relationships involves the more complicated business of *multivariate causation*. It often happens that human service systems come to recognize differences in the severity of the needs of their clients and to establish services of varying intensity to match the clients' needs. In such systems

Table 8.12
Illustrating the exposure of multivatiate causation

Bivariate analysis

		Outcome	
		Good	Bad
Intensity of service	Low	110 (61%)	70 (39%)
	High	130 (50%)	130 (50%)

Multivatiate analysis controlling for severity of need

Routine needs					Special needs			
		Outcome					Outcome	
		Good	Bad				Good	Bad
Intensity of service	Low	100 (91%)	10 (9%)		Intensity of service	Low	10 (14%)	60 (86%)
	High	80 (89%)	10 (11%)			High	50 (29%)	120 (71%)

Causal mechanism

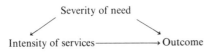

it is not uncommon to find that those clients receiving the highest intensity services have the worst outcomes, thereby casting doubt on the value of the more costly intensive services. The bivariate association between intensity of service and outcomes may mask the fact that the odds are stacked against those clients receiving intensive services, and a proper evaluation requires a kind of *severity adjustment*. The multivariate analysis will control for severity when examining the association between intensity of service and outcome, as in table 8.12. The table indicates that, although the rate of bad outcomes among those who receive intensive services is higher overall and slightly higher for routine cases who mistakenly receive intensive services (recall the discussion of sensitivity and specificity in chapter 2), it is lower among those clients with the greatest needs. Here severity of need and intensity of service both impact on outcome, and severity also impacts on intensity through whatever mechanism is established to identify clients with the greatest need and channel them to appropriate services. The general problem of severity adjustment is common throughout the evaluation of nonrandomized comparisons (see chapter 11) and in rate-setting processes that attempt to adjust reimbursement rates for variations in clientele across service providers.

These three illustrations indicate how adding a third attribute enriches an analysis. Statistically controlling for a third variable in this way is an imperfect but still helpful approach to building causal theories.

Summary

We can usually improve predictions by exploiting conditional probabilities. For discrete random variables it is natural to display the conditional distributions in the form of counts of cases arranged into a contingency table. When we use one attribute to predict another we ask two questions. First, how powerful is the predictor? We can document our answer to this question either by the proportional reduction in error that comes from using the predictor or by the proportional reduction in uncertainty. Second, is the degree of association between the attributes in the sample appreciably larger than that which would be created just by sampling artifacts when the two attributes are in fact independent in the population? We answer this second question by (1) determining how the counts would be distributed across the table if the null hypothesis of independence were true, (2) computing a measure of discrepancy between observed and expected counts for each cell of the table, (3) summing these to arrive at an overall measure of discrepancy for the table χ^2_{obs}, then (4) estimating the probability that sampling accidents alone could produce a discrepancy as large or larger if the null hypothesis were true. This probability is called the descriptive level of significance and serves as a reference by which to subjectively assess the credibility of the proposition that the predictive association really exists in the population. This assessment must weigh both the risk of overreacting to a sampling artifact and under-reacting to a real association.

We can usually further improve our predictions by considering more attributes, moving from bivariate to multivariate analyses. This tactic will often expose patterns of association that are hidden in bivariate analyses, and so provide help in suggesting or testing theories of relationships. Such phenomena as spurious correlation, causal intermediaries, and multivariate causation can be revealed by statistically controlling for additional attributes.

Nevertheless, causality can never be established solely on the basis of analysis of a single data set, especially if the data arise from an observational study rather than from an actual intervention. It is important that associations be replicable, explainable, and controllable before we treat them as causal. Furthermore, the advantages of adding another attribute to the analysis vanish when doing so will spread the counts too thin over the set of resulting tables or when the new attribute is itself highly associated with the others used for prediction (multicollinearity). Despite these cautions, analysis of multivariate contingency tables (and their continuous counterparts, multiple regression equations, see chapter 10) is very often the planner's best bet for mustering information in the face of uncertainty.

References and Readings

Baldus, D. and J. Cole. "Quantitative Proof of Intentional Discrimination." *Evaluation Quarterly* 1 (1977): 53–86.

Bishop, Y. M. M., S. E. Feinberg, and P. W. Holland. *Discrete Multivariate Analysis: Theory and Practice.* Cambridge, Mass.: The MIT Press, 1976.

Hoaglin, D. C. "Direct Approximations for Chi-Square Percentage Points." *Journal of the American Statistical Association* 72 (1977): 508–515.

Lindley, D. V. "The Bayesian Analysis of Contingency Tables." *Annals of Mathematical Statistics* 35 (1965): 1622–1634.

Mosteller and Rourke. "The Exact Chi-Square Distribution," chapter 8, and "Applying Chi-square: Two-Way Tables," chapter 11, pp. 141–158, 192–209.

Mosteller and Tukey. "Regression for Fitting," chapter 12, pp. 259–262.

Smith, M. "Almshouse Women: A Study of Two Hundred and Twenty-Eight Women in the City and County Almshouse of San Francisco." *Publications of the American Statistical Association*, 4 (1895): 219–262.

8.1

Compute the proportional reduction in error for the data in table 2.1, using first age and then activity limitation as the predictor.

8.2

Taggart studied the correlates of bank mortgage lending in 191 small areas in and around Boston and presented the tables that follow. Use the proportional reduction in uncertainty measure to compare the power of income data versus racial data for predicting bank-lending activity.

	Mortgage to sales percentage		
Mean household income	5–60	61–80	81–92
<$12,000	49	14	1
$12–16,000	9	25	32
>$16,000	1	23	37
Percentage of minorities in adult population	5–60	61–80	81–92
< 1	5	19	30
1–10	12	36	39
> 10	42	7	1

8.3

What is the descriptive level of significance of the association in the table of problem 8.2 relating mean household income in an area to bank-lending activity?

8.4

Okonjo's study of indigenous capital markets in rural Nigeria yielded the data that follows on a random sample of 50 men from the town of Ogwashi-Uku. The first digit of each triplet is 1 if the man had a second, part-time job and 0 otherwise. The second digit is 1 if the man used a commercial bank and 0 otherwise. The third digit is the number of indigenous savings and credit societies to which the man belonged. Use these data to test the argument: "The indigenous credit system serves as a substitute for the formal credit system, so greater involvement with the formal system implies lesser involvement with the indigenous system. In addition, to the extent that an individual can increase his income from part-time work, he will further decrease his participation in the indigenous system."

(0, 0, 5)	(0, 0, 4)	(0, 1, 5)	(1, 0, 3)	(1, 1, 4)
(0, 0, 2)	(0, 0, 7)	(0, 1, 5)	(1, 0, 4)	(1, 1, 6)
(0, 0, 0)	(0, 0, 10)	(0, 1, 3)	(1, 0, 5)	(1, 1, 5)
(0, 0, 6)	(0, 0, 4)	(0, 1, 3)	(1, 0, 2)	(1, 1, 3)
(0, 0, 8)	(0, 0, 1)	(1, 0, 7)	(1, 0, 1)	(1, 1, 3)
(0, 0, 3)	(0, 0, 5)	(1, 0, 0)	(1, 0, 2)	(1, 1, 0)
(0, 0, 3)	(0, 0, 0)	(1, 0, 4)	(1, 0, 3)	(1, 1, 3)
(0, 0, 4)	(0, 0, 0)	(1, 0, 5)	(1, 1, 4)	(1, 1, 5)
(0, 0, 1)	(0, 0, 0)	(1, 0, 4)	(1, 1, 7)	(1, 1, 1)
(0, 0, 3)	(0, 1, 4)	(1, 0, 8)	(1, 1, 7)	(1, 1, 10)

9 Bivariate Regression Analysis: Conditional Prediction of a Continuous Variable from a Single Attribute

In chapter 8 we considered using one (or more) attributes to predict a variable that takes on one of a discrete set of values. Although the techniques of contingency table analysis can be used for predicting either nominal, ordinal, or metric variables, we can generally make better use of continuous metric data by switching from analysis of contingency tables to the techniques of regression analysis described in this chapter and the next. Just as the contingency table offered a way to move from an unconditional to a conditional prediction of the modal category of the uncertain discrete attribute, so regression analysis offers a way to move from an unconditional to a conditional prediction of the mean value of a continuous attribute. Both contingency tables and regression analyses allow for systematic adjustment of predictions based on auxiliary information. In this chapter and the next we will encounter several notions familiar from analysis of contingency tables: errors in prediction, measures of strength of association, multicollinearity, significance tests, and questions of causality. We will also encounter some new notions: selecting values for the intercept and slope of a straight line and transforming data so that they plot closer to a straight line.

In this chapter we will again make use of the concept of highest density region (HDR), last used in chapter 7 on estimating a population mean. In fact, this chapter and the next should really be thought of as a merger of the previous two, since we return to the issue of estimating a mean, only now we make use of conditioning information to improve the estimate.

Recall the data shown in table 7.1 that sampled the full-value tax rates in 11 Massachusetts communities. The sample mean of $27.42 was offered as a simple, but not unreasonable, prediction of the tax rate in other Massachusetts communities. However, we should be able to improve our predictions if we condition them on some attribute of the city, perhaps age or population or level of industrial development. How to make these conditional predictions and how to assess them are the subjects of our study of regression analysis.

Regression as Curve Fitting

There is a well-developed theory of statistical inference for regression which we will take up in a while. First, we view regression analysis as a descriptive methodology for fitting a straight line to a set of points. Such a set of points is called a *scattergram* or scatter diagram. It displays the association between two attributes of the cases. Our goal in regression analysis is to summarize the relationship between the two attributes in a simple formula that can then be used to predict the value of one attribute from knowledge of the other.

Such a scattergram is shown in figure 9.1, which plots tax rate and population (in thousands) for the sample of 11 Massachusetts communities. Examination of the points indicates a positive association between population (POP) and full-value tax rate (TAX): the larger the population, the higher the tax rate. It seems that if someone asks us to estimate the tax rate in some other Massachusetts community—say, South Hadley—it would be worthwhile to find out the population before we make our prediction. If the population of South Hadley is large, we should edge our estimate up somewhat.

Shown in figure 9.1 is the *regression line*:

$$\widehat{TAX} = \$21.66 + \$0.72POP.$$

(The carat above TAX signifies a predicted quantity.) This line gives a "rule" relating the population of a town to our prediction of its tax rate. The rule says to start with a base of $21.66 and then add $0.72 to our prediction for every 1,000 residents. It happens that South Hadley has a population of 16,568, so our rule says to predict a tax rate of

$$\widehat{TAX} = \$21.66 + \$0.72(16.568) = \$33.59$$

for South Hadley. If we were to ignore the population, our best prediction would be $27.42, the mean tax rate in the sample of 11 cities. It happens that the tax rate in South Hadley is $34.41, so (in this case, at least), it pays to use the regression prediction rather than the sample mean: investing in the population datum reduced our prediction error from $34.41 − $27.42 = $6.99 to $34.41 − $33.59 = $0.82.

Measures of the Predictive Power of the Regression Line

Examining figure 9.1, we can see that using the regression line rather than the sample mean gives a better estimate in 6 of the 11 cases: Savoy, Monterey, Belchertown, Northborough, Easthampton, and Wellesley.

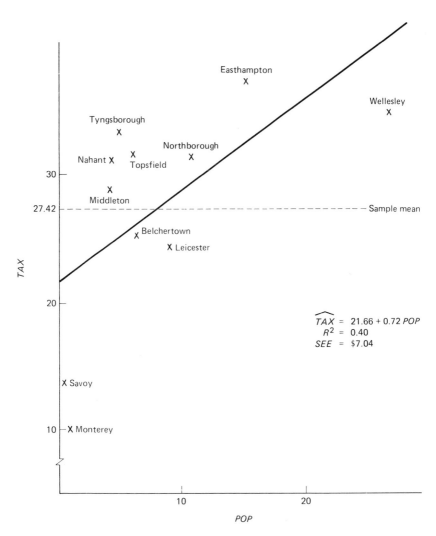

Figure 9.1
Straight line fit relating population in 1,000's (*POP*) to full-value tax rate per $1,000 value (*TAX*)

How else might we objectively summarize the *goodness of fit* of the regression line to the data points?

There are two indices of goodness of fit in common use. Both are based on the *residuals* or errors in prediction (the residuals for South Hadley are \$0.82 when predicting from the regression line and \$6.99 when predicting from the sample mean). The first index, called R^2 (read "*R*-square") or the *coefficient of determination*, compares the residuals from the regression predictions to the residuals from the sample mean. The formula for R^2 is analogous to those for λ or U. Instead of working with counts of categorical prediction errors (as in λ) or the entropy of the discrete distribution of the predicted variable (as in U), we work with the sum of squared residuals as the basis for assessing predictive power. Squaring the residuals insures that underestimates (positive residuals) will not cancel out overestimates (negative residuals); it also gives extra attention to the largest errors in prediction. The proportional reduction in the sum of squared residuals obtained by exploiting the attribute data is called R^2 and evaluates the regression estimate relative to the simpler estimate provided by the sample mean alone:

$$R^2 = \frac{\begin{array}{c}\text{Sum of squared residuals} \\ \text{from sample mean}\end{array} - \begin{array}{c}\text{Sum of squared residuals} \\ \text{from regression line}\end{array}}{\text{Sum of squared residuals from sample mean}}.$$

To put the definition formulaically, suppose we want to use attribute X to predict attribute Y using sample data. Let

$Y_i =$ value of the variable to be predicted for the ith case in the sample; Y_i is called the "predicted" variable,

$X_i =$ value of the attribute on which the prediction will be based for the ith case in the sample; X_i is called the "predicting" variable,

$\bar{Y} =$ the sample mean of the predicted variable,

$\hat{Y}_i =$ the value predicted for the ith case using the regression rule,

$n =$ the sample size.

Now the residual or prediction error for the ith case using the sample mean $\bar{Y}$ as the prediction is $(Y_i - \bar{Y})$, and the residual using the regression estimate is $(Y_i - \hat{Y}_i)$. Therefore

$$R^2 = \frac{\sum_{i=1}^{n} (Y_i - \bar{Y})^2 - \sum_{i=1}^{n} (Y_i - \hat{Y}_i)^2}{\sum_{i=1}^{n} (Y_i - \bar{Y})^2} .$$

The value of R^2 ranges from 0 to 1.0, depending on how well one can predict Y using the attribute X relative to predictions made ignoring the attribute. If the attribute X makes no difference to the predictions, then $\hat{Y}_i = \bar{Y}$ and $R^2 = 0$. Note too that if the attribute X allows perfect prediction, so that $Y_i = \hat{Y}_i$ exactly, then $R^2 = 1.0$. The regression predicting tax rate from population in figure 9.1 has $R^2 = 0.40$. This is clearly an improvement over ignoring the population attribute and using just the sample mean as a prediction.

The second commonly used measure of goodness of fit is the *standard error of estimate* (*SEE*). This measure is very similar to the sample standard deviation. Whereas the sample standard deviation is a measure of the typical size of the residuals when predictions are based on the sample mean alone, the standard error of estimate is a measure of the typical size of residuals when the regression prediction is used. For prediction based on k attributes (in this chapter $k = 1$), the standard error of estimate is computed as

$$SEE = \sqrt{\frac{1}{n - k - 1} \sum_{i=1}^{n} (Y_i - \hat{Y}_i)^2} .$$

The *SEE* for the regression in figure 9.1 is $7.04, meaning that predicting tax rate from population for the cases in the sample gave estimates that were typically in error by about $7.04. For some problems the absolute size of the error is important (we may have to make predictions with typical errors less than $1.00 to satisfy our purposes), and the *SEE* can be used directly as a measure of goodness of fit. If the proportional size of the error is important, we can gauge our reaction to the *SEE* by comparing it to that typical value of the predicted variable known as the mean; in such cases it is instructive to compute the ratio $SEE / \bar{Y}$, which conveys an impression of the percentage error in prediction. For the tax rate example

$SEE / \bar{Y} = \$7.04/\$27.42 = 0.26$,

so the typical error in prediction is around 26 percent. This represents a slight improvement over the coefficient of variation for the sample, which, since the sample standard deviation is $8.59, equals $8.59/$27.42 = 0.31.

Choosing the Best-Fitting Straight Line

We have not yet said how it happened that the particular straight line $\widehat{TAX} = \$21.66 + \$0.72POP$ got to be our regression rule in the tax rate example. Looking at the data points in figure 9.1 and fitting a line by eye might get us any number of different straight lines that seem close to the data points—or to most of them. In part, it is the subjectivity and variability that characterize such informal eyeballing methods that make an objectively computed line attractive. We can avoid arguments about which line fits best by specifying an algorithm for computing the line, and we can let the computer do the fitting.

We must have a criterion of fit. Several are possible. The one most commonly used is the *least-squares* criterion: pick that straight line that minimizes the sum of squared residuals about the line. This is equivalent to picking that straight line that maximizes the coefficient of determination R^2. Now a straight line is characterized by two numbers: a slope and an intercept. Once we pick an intercept and a slope we have created a candidate regression line. We can imagine generating a large number of such candidates, computing from each the values $\hat{Y}_i$ of the predicted variable, summarizing the fit by R^2, then retaining that one line that produces the highest R^2. (Incidently but not accidently, we would get the same result if instead of maximizing R^2 we were to minimize SEE; both goals are equivalent to minimizing the sum of squared residuals.)

It happens that the methods of calculus can be used to derive equations for computing from the data points that particular slope and intercept which give the best-fitting straight line (as measured by R^2 or SEE). It is customary to use the notation B_0 to designate the intercept and B_1 to designate the slope. The general form of the regression equation for predicting the value of Y_i from knowledge of the attribute X_i is then

$\hat{Y}_i = B_0 + B_1 X_i$.

The formulas for the slope and intercept are

$$B_1 = \frac{\sum\limits_{i=1}^{n} (X_i - \bar{X})(Y_i - \bar{Y})}{\sum\limits_{i=1}^{n} (X_i - \bar{X})^2}$$

$$B_0 = \bar{Y} - B_1\bar{X},$$

where $\bar{X}$ and $\bar{Y}$ are the sample means.

Once the computer has calculated the slope and intercept of the best-fitting straight line, we have a handy way to summarize the information contained in the scattergram: rather than toting around the entire scattergram, we can tote around the more portable regression equation.

The Correlation Coefficient

Often we want to determine the regression line in order to predict the value of the Y variable based on knowledge of the value of the X variable. Sometimes, however, all we want to know is the general nature of the association between the two variables, and we do not care to know the actual value of the Y variable. The issue becomes more general: are higher values of X associated with higher or lower values of Y? In such circumstances we could further reduce our summary of the scattergram by ignoring the intercept and focusing only on the slope of the regression line. While such a single number summary might suffice, we can do a little better by computing a different, but related, single number summary: the *correlation coefficient*, denoted by R.

If the actual values of the X and Y variables are unimportant to us, we should feel free to work with their standardized values, which we will call Z_x and Z_y. Standardizing the data and then finding the least squares fit produces a regression line that predicts the standardized value of Y from the standardized value of X:

$$\hat{Z}_y = RZ_x.$$

(A least squares fit to standardized variables always produces a line that passes through the origin, which accounts for the lack of an intercept in the resulting equation.) The slope R is proportional to the slope B_1 we would have obtained without standardizing, so it will have the same sign, but it carries extra information in its magnitude since its square happens to be the coefficient of determination R^2, which measures goodness of fit.

The value of R will range from -1.0 to $+1.0$. A value $R = -1.0$ indicates a perfect inverse linear relationship between the two variables (higher X goes with lower Y), and a value $R = +1.0$ indicates a perfect direct linear relationship (higher X goes with higher Y). For the tax and population data in figure 9.1, $R = +0.63$.

When searching for bivariate relationships in large data bases, you may want to use the correlation coefficient as a kind of mass screening device: if you have, say, 12 variables of interest in a problem, you can use the computer to produce a table of all $\binom{12}{2} = 66$ pairs of correlation coefficients. You can then scan the table to spot interesting relationships for further detailed investigation. Please note, though, that the only advantage in this procedures is that it saves paper compared to plotting all 66 scattergrams. There are ways to fool a correlation coefficient (for instance, data forming a rainbow-shaped pattern in a scattergram will produce a very small value of R even though there is a clear, albeit non-linear, association). It is always best to eyeball the data if you possibly can. The correlation coefficient, coefficient of determination and standard error of estimate will serve as useful numerical summaries around which to build your interpretation of the patterns you see in the data (if any), but you should never underestimate the importance of the scattergram as a tool for bivariate data analysis.

Data Transformations

There are real advantages to fitting straight lines to data: straight lines are easy to describe (slope and intercept), the eye seems especially sensitive to departures from linearity, and most computer programs for regression analysis are designed to fit straight lines only. Unfortunately, not all scattergrams will show an essentially linear relationship between two variables, and you may rightly feel that fitting a straight line to a particular set of data points is unrealistic (your visual impression of poor fit will receive confirmation from a relatively low value of R^2). Fortunately, there is often a way to retain the advantages of straight line fits without straining the credibility of the analysis. The trick is to work not with the data as given but rather to alter systematically the vertical and / or horizontal spacing of the data points by the application of a mathematical transformation.

Recall that when selecting the best-fitting straight line the unknowns are the intercept and slope of the regression line, not the data points themselves. Since the pairs of values (X_i, Y_i) are known, we can readily compute transformed pairs, such as $(\sqrt{X_i}, \sqrt{Y_i})$, (X_i^2, Y_i^3), $(X_i, \log Y_i)$, or $(1/X_i, \exp[Y_i])$, and fit a straight line to the transformed data. It often happens that the transformed data are better fit by a straight line than are the original data. Then if, for instance, we can compute a good estimate of $\sqrt{Y}$, we can reverse the transformation and use as our predicted value of Y the square of the predicted value of $\sqrt{Y}$. Such a procedure amounts to fitting a curved line to the original data, but we actually determine the regression line by fitting a straight line to the transformed data.

To see the value of a very commonly used transformation, return to figure 9.1 and note that the regression line is rather far away from the data points for the tiny communities of Savoy and Monterey. Now skip ahead to figure 9.4, and see how much better a fit is provided by a curved line that rises steeply from the lower left corner and gradually bends over as population increases. This particular curve is called a *power curve* and will serve to illustrate the process of data transformation.

Working with data in their original (X, Y) form, we fit the straight line $\hat{Y} = B_0 + B_1 X$. Suppose instead that we work with the transformed data $(\log X, \log Y)$; then we obtain a prediction not of Y from X but of $\log Y$ from $\log X$:

$$\widehat{\log Y} = B_0 + B_1 (\log X).$$

Now if we raise 10 to the $\widehat{\log Y}$ power, we undo the logarithmic transformation and obtain a prediction for Y itself:

$$\begin{aligned}
\hat{Y} &= 10^{\widehat{\log Y}} \\
&= 10^{(B_0 + B_1 \log X)} \\
&= 10^{B_0} \times 10^{B_1 \log X} \\
&= 10^{B_0} \times 10^{\log X^{B_1}} \\
&= 10^{B_0} \times X^{B_1}.
\end{aligned}$$

The term 10^{B_0} is easy to compute since B_0 is a value provided by the regression analysis which predicted log Y from log X; likewise B_1 is provided by the same analysis. Therefore we have a rule which ultimately converts a value of X into a prediction $\hat{Y}$. Since the prediction involves raising X to the B_1 power, this nonlinear prediction rule is known as a *power curve*. For obvious reasons, the data transformation is known as a *log-log transformation*.

The key parameter in the power curve is the power itself, B_1. Different numerical values of B_1 correspond to qualitatively different curves, as shown in figure 9.2. For values in the range $B_1 > 1.0$, the power curve increases at an increasing rate. For values in the range $0 < B < 1.0$,

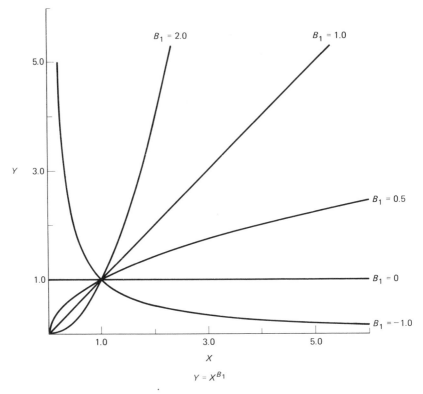

Figure 9.2
Power curves

the power curve increases at a decreasing rate. For values in the range $B < 0$, the power curve decreases. Thus a rather wide range of behaviors can be produced by the same functional form. Incidently, the power curve plays a prominent role in econometric studies. If the variable Y represents the total cost of producing a good and X represents the total volume produced, then values of $B_1 > 1.0$ indicate diseconomies of scale in production, while values in the range $0 < B_1 < 1.0$ indicate economies of scale. In econometric studies the quantity B_1 is known as the *elasticity* of Y with respect to X; it can be interpreted as the percentage change in Y associated with a 1.0 percent change in X.

Returning to the tax rate example, the log-log transformation is illustrated in figure 9.3. Contrast this improved straight line fit with the fit to the original, untransformed data in figure 9.1. After the log-log transformation, the data points for Savoy and Monterey look much more "natural," since even though they are isolated from the main body of points they fall reasonably close to a straight line extended from that main body. A scattergram like figure 9.3 can be had either by computing the logs of population and tax rate and plotting the transformed data on regular (linear) graph paper or by plotting the original data on log-log graph paper.

If you wish to compare directly the predictive ability of the power curve relative to the original straight line fit, it is not proper to compare the R^2 values of figures 9.1 and 9.3. This is because figure 9.3 illustrates the ability to predict not the tax rate itself but rather its logarithm. To make a proper comparison you must use the log-log fit to produce power curve predictions of the tax rate itself and then determine the residuals directly. Since the log-log fit, provided the regression prediction is

$$\widehat{\log TAX} = 1.20 + 0.30 \log POP,$$

the power curve fit is

$$\widehat{TAX} = 10^{1.20} \, POP^{0.30}$$
$$= 15.97 \, POP^{0.30}.$$

This curve is shown in figure 9.4. Using the power curve to predict full-value tax rate from population gives an $R^2 = 0.63$ and a $SEE = \$5.47$, both improvements over the original straight line fit with $R^2 = 0.40$ and $SEE = \$7.04$. Note especially that these improvements in fit came

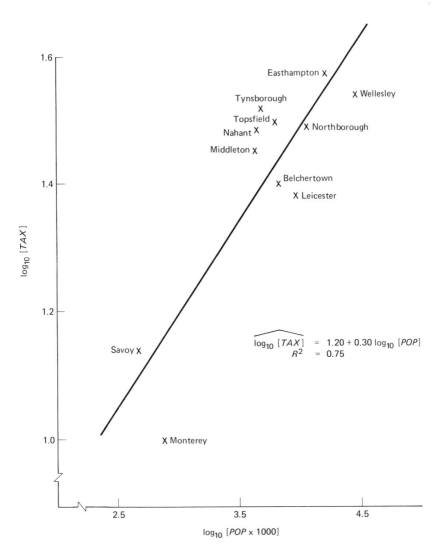

Figure 9.3
Straight line fit to transformed data

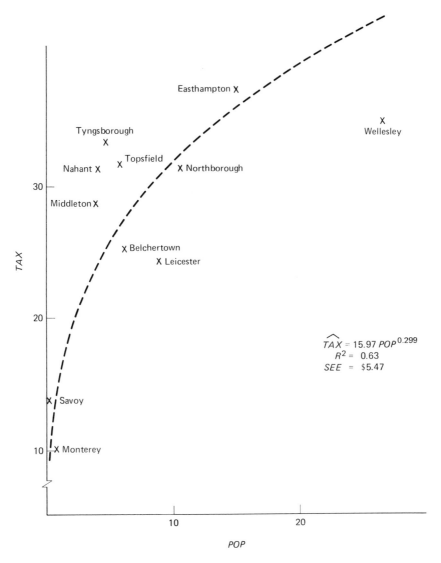

Figure 9.4
Power curve fit relating population in 1,000's (*POP*) to full-value tax rate per $1,000 value (*TAX*)

about without adding either more predicting variables or more data points.

Although the log-log transformation is one of the most commonly used, many other transformations are helpful in practice. Tufte's book contains good illustrations of the use of logarithmic transformations; Mosteller and Tukey's text explains a simple procedure for determining which of a broad class of transformations will do best at linearizing a given set of data.

Highest Density Regions for Predictions

When we made unconditional estimates of a mean in chapter 7, we found it useful to make our uncertainty explicit by citing highest density regions. Regression analysis allows us to condition our predictions on one (or more) attributes; but even though exploiting the conditioning information will improve our predictions, we will still feel the need for some sense of the uncertainty involved. For a quick indication, we can use the standard error of estimate, but we can go further and establish highest density regions (HDR) for the predicted values, using essentially the same techniques as we used in chapter 7 when making unconditional estimates of a mean value.

Suppose we have a sample of cases from a much larger population which provides the pairs of (X, Y) values. We assume that the mean value of Y changes linearly with the value of X and that at any particular value of X the value of Y has a Gaussian distribution. That is, if we happened to have many cases with the same value of X, their Y values would have a Gaussian distribution centered on a mean that depends linearly on X. We further assume that although the mean value of Y varies with X, the standard deviation of Y stays constant across the full range of X. Finally, we assume that the fluctuations of the Y values at different values of X are independent. When these four assumptions are met, we can bring into play the machinery for computing highest density regions.

Let

B_0 = intercept of regression line
B_1 = slope of regression line
n = sample size

$\bar{X}$ = sample mean of X

S_X = sample standard deviation of X

SEE = standard error of estimate of regression line;

$$= \sqrt{\frac{\sum_{i=1}^{n} [Y_i - (B_0 + B_1 X_i)]^2}{n - 2}}.$$

Then a 95 percent HDR for the mean value of Y conditional on the fact that $X = X_0$ (and assuming a flat prior) is

95 percent HDR for mean Y

$$= (B_0 + B_1 X_0) \pm C_{n-2} \times SEE \times \sqrt{\frac{1}{n} + \frac{1}{(n - 1)} \left(\frac{X_0 - \bar{X}}{S_X}\right)^2},$$

where C_{n-2} is the 97.5 percent point of Student's t distribution with $n - 2$ degrees of freedom.

Note that the HDR is centered on the value predicted by the regression line. Note too that the width of the HDR depends on the standardized value of the distance between the value of X at which the prediction is made and the mean value of X in the sample: $\dfrac{X_0 - \bar{X}}{S_X}$. The farther away from the mean, the wider the HDR. Put another way, when we try to make predictions at values of X way out at the very edge of our empirical experience, we must recognize the greater uncertainty inherent in our predictions. For instance, a prediction made from a value of X that lies 2 sample standard deviations out from the typical (mean) value of X will have a HDR that is about $\sqrt{5} = 2.24$ times as wide as the HDR for a prediction made from a safe place in the middle of the data at the sample mean $\bar{X}$. If we want to be more confident about our predictions at such extreme values of X, then we must simply acquire more experience (get more data) at the edges. You should therefore note the value of seeking out for analysis cases having very low or very high values of the predictor variable X.

Note that the formula gives a 95 percent HDR for the mean value of Y conditional on the value of X. It is quite another matter to construct a 95 percent HDR for the predicted value of Y conditional on X for an individual case. We have learned already that the mean is a less volatile quantity than the individual values from which it is computed, so we expect wider HDR's for predictions of individual values than for predictions of means. In fact, the 95 percent HDR for an individual value of Y conditional on the fact that $X = X_0$ is

95 percent HDR for single Y

$$= (B_0 + B_1 X_0) \pm C_{n-2} \times SEE \times \sqrt{1 + \frac{1}{n} + \frac{1}{(n-1)} \left(\frac{X_0 - \bar{X}}{S_X}\right)^2}.$$

In this case the square root includes the additional term 1.0, which increases the width of the HDR for an individual value relative to that for a mean. As before, the width of the HDR depends on how close the value X_0 is to the frontier of experience represented in the data; the width (and therefore the uncertainty in the prediction) is least when predicting from the mean of X.

For a numerical example, consider again the prediction of tax rate from population. In the 11 communities the sample mean of the population (in thousands) was $\bar{X} = 7.98$, and the sample standard deviation was $S_X = 7.84$. In the regression using the untransformed data (figure 9.1) the prediction rule was

$$\widehat{TAX} = 21.66 + 0.72 POP$$
$$R^2 = 0.40$$
$$SEE = \$7.04.$$

Therefore, for a community with a population of 10,000 people, the predicted full-value tax rate is

$$\widehat{TAX} = 21.66 + 0.72(10) = \$28.86.$$

The 95 percent HDR for the mean tax rate of many communities of 10,000 is

95 percent HDR for mean of cities of 10,000

$$= \$28.86 \pm C_9 \times SEE \times \sqrt{\frac{1}{11} + \frac{1}{10} \left(\frac{10 - 7.98}{7.48}\right)^2}$$
$$= \$28.86 \pm 2.262 \times 7.04 \times \sqrt{0.09}$$
$$= \$28.86 \pm \$4.99.$$

In contrast, if we are attempting to predict not the mean tax rate for a number of cities of 10,000 population but rather the rate for a single city, then

95 percent *HDR* for single city of 10,000

$$= \$28.86 \pm C_9 \times SEE \times \sqrt{1 + \frac{1}{11} + \frac{1}{10}\left(\frac{10 - 7.98}{7.48}\right)^2}$$

$$= \$28.86 \pm \$16.69.$$

The prediction for an individual city is much less certain than that for the mean of many cities with the same population. Finally, note how the uncertainty increases when we try to predict the mean tax rate among cities of 20,000 population, which is a size larger than all but one of the communities in the sample:

95 percent *HDR* for mean of cities of 20,000

$$= \$36.06 \pm C_9 \times SEE \times \sqrt{\frac{1}{11} + \frac{1}{10}\left(\frac{20 - 7.98}{7.48}\right)^2}$$

$$= \$36.06 \pm \$9.41.$$

There is great uncertainty associated with this prediction—as there should be, since the prediction is based on a rather unimpressive fit to a small sample of communities and is made for cities of a size not well represented in the data. If the data had contained more cities of the appropriate size and the relationship between population and tax rate had been more nearly linear, then the uncertainty in the prediction would have been considerably reduced. Keeping track of the uncertainty by means of highest density regions helps a good deal to keep us appraised of the limitations of our fit.

Crossvalidation

It is not uncommon (but often unfortunate) for a planner to discover that a regression line that does well at summarizing the relationship between two attributes in one sample turns out to predict rather poorly when used with new data. The process of picking the intercept and slope that give the least-squares fit to a particular data set tends to emphasize idiosyncratic features of that data that may not be present in other samples. Therefore the same caution applies to regression analysis as to analysis of contingency tables: confidence in the validity of an association is best strengthened by repeating the analysis on data sets different from that which first revealed the association. This replication, or crossvalidation, process is essential to developing a sense of the reliability of a prediction rule in practice settings different from that in which the rule was first constructed.

While it is usually more valuable to replicate the analysis with different data sets, one usually begins with only a single set of data. Nevertheless, even so early in the analysis process it is usually possible to perform a crossvalidation by holding some data in reserve. Deriving a regression line from, say, half the data and testing its predictive power on the remaining half, then repeating the procedure in reverse, is more work than simply deriving a single regression line from the full set of data but leads to a more realistic assessment of predictive power. My personal rule of thumb in regression analysis is to try to have at least 10 data points for every predictor variable, so the "half and half" strategy for crossvalidation should work quite comfortably with data sets having 20 or more cases. Incidently, the performance of the regression rule need not always deteriorate when applied to data different from those by which it was determined: when I applied the power curve $\widehat{TAX} = \$15.97POP^{0.30}$ to a new set of 11 Massachusetts communities averaging 3 times the population of the original set, the predictive rule actually worked slightly better ($R^2 = 0.69$, $SEE = \$5.25$).

Summary

When predicting the value of a continuous metric attribute, we can improve our predictions by exploiting the conditioning information available in a second such attribute. The association between the two attributes is summarized in the values of the slope and intercept of a straight line fitted to the data. By far the most common method of fitting chooses the slope and intercept in such a way as to minimize the sum of squared errors in prediction. The predictive power of the resulting regression line is usually summarized in absolute terms by the standard error of estimate, which describes the typical size of a prediction error. Since the sample mean provides an alternative, simpler prediction, the utility of the regression line is also usually summarized by the coefficient of determination R^2, which (analogous to λ for contingency tables) indicates the proportional reduction in the sum of squared residuals achieved by using the regression line instead of the sample mean. When we care only to know the direction and strength of linear association between two metric attributes, we may summarize the relationship using the correlation coefficient R, which can be thought of as the slope of the regression line fitted to the standardized version of the data.

When the original data displays a distinct pattern when plotted in a scattergram but the pattern is poorly fit by a straight line, it often helps to transform the data before fitting a regression line. The log-log transformation is one important transformation; fitting a straight line to the transformed data amounts in this case to fitting a power curve to the original data.

Whenever we make predictions, we want to have some assessment of the uncertainty involved, and highest density regions can provide this information for regression predictions just as they did for estimates of a population mean. The highest density regions will be wider for predictions of individual values than for predictions of means, and they will be wider when working far from the mean value of the predicting variable. As with contingency tables, you should not make too much of the analysis of one sample but should replicate the analysis on other data sets to document its validity. A good start at crossvalidation involves dividing the original data set into one piece that is used to calculate the regression line and a second piece that is used to test the predictive power of the line so derived.

References and Readings

Eaton, I. "Receipts and Expenditures of Certain Wage-Earners in the Garment Trades," *Publications of the American Statistical Association* 4 (1895): 135–180.

Mosteller and Tukey. "Indications of Quality: Cross-Validation," section 2.F, and "Straightening Curves and Plots," chapter 4, pp. 36–40, 79–88.

Phares, D. "Property Taxation and Equity: An Interstate Analysis." *Annals of Regional Science* 7 (1973): 27–39.

Schmitt. "Straight Line Analysis," section 9.3, pp. 290–308.

Tukey. "Straightening Out Plots (Using Three Points)," chapter 6, pp. 169–203.

9.1

In problem 3.1 you were asked to use the sample mean (or median) to estimate missing values in McClure's data on Cambridge neighborhoods. (a) Determine a new estimate for the mean household income in neighborhood 12 using one of the 5 known values of other attributes of neighborhood 12 and determine the 95 percent *HDR* for the estimate. (b) Compare the 95 percent *HDR* to that obtained using only the sample values of mean neighborhood income.

9.2

An important issue in data analysis is the choice of unit of analysis, the level of aggregation of the data. Explore this issue by using the following data set on crime rates in Florida to compare results at the city level to results at the SMSA level. In each case use the data to compute the entropy of the distribution of family income, then predict crime rate from income mix as reflected in the entropy measure. Compare the two analyses.

Area	Percentage of families with 1969 income < poverty level		Percentage of families with 1969 income < 1.25 poverty level		Serious crimes reported to police per 100,000 residents (1975)	
	City	SMSA	City	SMSA	City	SMSA
Boca Raton	4.6	10.3	7.3	14.9	5951	8932
Bradenton	17.8	13.4	26.0	21.1	6317	6016
Gainsville	14.0	15.4	18.3	20.4	8997	8518
Miami	16.4	10.9	22.8	15.4	10910	9007
Orlando	15.3	11.6	21.4	16.8	10854	8398
Panama City	17.3	14.9	25.3	22.7	7568	5630
Pensacola	17.1	15.5	22.3	21.4	10215	7644
Sarasota	12.8	11.7	18.4	16.8	9006	6702
Tampa-St. Petersburg	10.8	11.3	16.5	16.8	8613	7596
Tallahassee	13.3	14.5	17.6	18.9	8079	7714

9.3

In his review of the Housing and Community Development Act of 1974, R. L. Cole analyzed the following data on the increase in funds provided by the Act compared to the previous categorical grant programs. One of Cole's conclusions was that smaller, western SMSA's were receiving disproportionate increases compared to larger northeastern SMSA's. Cole singled out as evidence the percentage differences in Dallas, Fort Worth, Phoenix, Omaha, El Paso, New York, Philadelphia, Baltimore, Boston, Pittsburgh, Buffalo, and Newark. Use regression analysis to predict the percentage difference in receipts in these SMSA's from the previous level of receipts. Do the residuals for these SMSA's support Cole's claim that geography was an important factor?

Community development entitlement amounts for 50 largest cities in 1975 (thousands of dollars)

City	Central city entitlements	SMSA entitlements	Previous SMSA Receipts[a]	Percentage difference
New York	$102,244	$122,042	$121,213	.68
Chicago	43,201	51,686	45,005	14.84
Los Angeles	38,595	71,592	67,617	5.87
Philadelphia	60,829	71,238	68,425	4.11
Detroit	34,187	58,536	54,390	7.62
Houston	13,257	15,218	13,523	12.53
Baltimore	32,749	35,302	33,917	4.08
Dallas	3,998	13,191	9,149	44.18
Washington, D.C.	42,748	42,748	42,748	0.00
Cleveland	16,092	20,733	18,981	9.23
Indianapolis	13,929	14,466	13,929	3.86
Milwaukee	13,383	14,224	13,486	5.48
San Francisco	28,798	58,839	55,375	6.26
San Diego	9,148	12,200	10,491	16.29
San Antonio	17,904	18,466	18,171	1.62
Boston	32,108	52,541	50,787	3.45
Memphis	6,043	6,344	6,043	4.98
St. Louis	15,194	20,015	17,908	11.77
New Orleans	14,808	16,493	14,930	10.47
Phoenix	2,570	5,493	4,018	47.96
Columbus	9,194	10,178	9,194	10.70
Seattle	11,641	14,303	13,265	7.83
Jacksonville	5,193	5,400	5,193	3.99
Pittsburgh	16,429	36,282	33.466	8.41
Denver	15,805	18,291	17,293	5.77

Community development entitlement amounts (continued)

City	Central city entitlements	SMSA entitlements	Previous SMSA Receipts[a]	Percentage difference
Kansas City, Mo.	17,859	20,074	19,799	1.39
Atlanta	18,780	22,228	20,670	7.53
Buffalo	11,685	17,209	15,857	8.53
Cincinnati	18,828	21,918	20,964	4.55
Nashville-Davidson	9,609	11,899	11,507	3.41
San Jose	6,554	8,505	7,089	19.97
Minneapolis	16,793	39,622	38,147	3.87
Fort Worth	1,879	13,191	9,149	44.18
Toledo	11,831	12,123	11,831	2.47
Newark	20,513	27,288	26,217	4.09
Portland	8,760	9,310	8,760	6.28
Oklahoma City	8,183	10,459	9,937	5.25
Louisville	8,639	11,648	11,496	1.32
Oakland	12,738	58,839	55,375	6.26
Long Beach	1,514	51,686	45,005	5.87
Omaha	1,390	1,496	794	88.41
Miami	3,165	25,851	24,983	3.47
Tulsa	9,312	10,219	9,805	4.22
Honolulu	13,099	13,099	13,099	0.00
El Paso	2,195	2,295	854	168.74
St. Paul	18,835	39,622	38,147	3.87
Norfolk	17,766	23,705	22,993	3.10
Birmingham	5,040	7,980	6,326	26.14
Rochester	14,684	17,911	17,304	3.51
Tampa	8.577	11,747	9,410	24.84

[a] Includes all money received from the consolidated categorical grant program.

9.4

Hadaway studied the rates of return on agricultural land investments between 1954 and 1974 in 10 regions of the United States and reported the following correlation coefficients among regional rates of return. Which region is generally most attractive from the perspective of diversification of a portfolio of agricultural land investments? (Key: NE = northeast, LK = lake states, CB = corn belt, NP = northern plains, AP = appalachia, SE = southeast, DS = delta states, SP = southern plains, MT = mountain, PA = pacific.)

Correlation of returns among regions (1955 to 1974)

	NE	LK	CB	NP	AP	SE	DS	SP	MT	PA	US
NE	1.00	0.883	0.822	0.824	0.889	0.786	0.706	0.773	0.837	0.203	0.859
LK		1.00	0.872	0.841	0.892	0.862	0.712	0.662	0.847	0.427	0.902
CB			1.00	0.920	0.880	0.811	0.857	0.738	0.856	0.547	0.958
NP				1.00	0.916	0.786	0.784	0.722	0.890	0.525	0.953
AP					1.00	0.825	0.769	0.805	0.930	0.427	0.953
SE						1.00	0.778	0.689	0.813	0.646	0.888
DS							1.00	0.632	0.739	0.606	0.859
SP								1.00	0.761	0.401	0.796
MT									1.00	0.480	0.937
PA										1.00	0.580
US											1.00

10 Multivariate Regression Analysis: Conditional Prediction of a Continuous Variable from Several Attributes

Having used one attribute to predict another, we naturally wonder whether adding yet another attribute would improve our predictions. When our data were counts arranged in contingency tables (chapter 8), introducing a third attribute meant subdividing the original table into as many new tables as there were categories of the new attribute (for instance, making separate tables of mortgage lending by income for minority and non-minority neighborhoods). When doing regression analysis, adding a new attribute means trying to predict the residuals left over from the previous analysis. In fact, multivariate regression can be thought of as a sequence of bivariate regressions, each performed on the residuals produced by the previous stage of analysis.

Multivariate Regression as the Sequential Analysis of Residuals

Residuals are the key to regression analysis. We have seen that the sum of squared residuals usually provides the criterion for choice of the intercept and slope of the bivariate regression line; it also forms the basis of the most commonly used measures of predictive power (R^2 and *SEE*). More importantly, however, the residuals are our prediction errors and therefore may tell us something about how to improve our forecasts by using additional attributes.

A regression line summarizes in concise form a pattern discovered in a data set. A case whose value is well predicted is behaving typically, while a case with a large residual is somehow unusual and a proper target for further analysis. The cycle of summary, analysis of residuals, re-summary,

Table 10.1
Data for a hypothetical example of multivariate regression

Case number	X	Y	Z
1	5	8	6
2	11	1	15
3	12	21	7
4	19	5	19
5	25	24	15
Sample mean	14.40	11.80	12.40
Sample standard deviation	7.73	10.13	5.64

re-analysis of residuals, and so on, comprises the process for extracting the information available in a data set. Of course, the cycle cannot repeat too often before we reach a point of overanalyzing the data, either because the summary is barely smaller than the data set itself or because the remaining prediction errors are of trivial size and uninteresting origin. Still, we can often do much better by repeating the cycle a few times than by stopping after only one pass through, so most regression analyses done in planning contexts involve more than one predictor variable.

To see how this cycle of successive analyses of residuals works, consider a small contrived example (we will follow with a more realistic example after the basic notions are clear). Our goal is to use attributes X and Y to predict attribute Z; data for the example are given in table 10.1. We begin by using only X to predict Z. The scattergram in figure 10.1 shows a direct relationship, with larger values of X generally predictive of larger values of

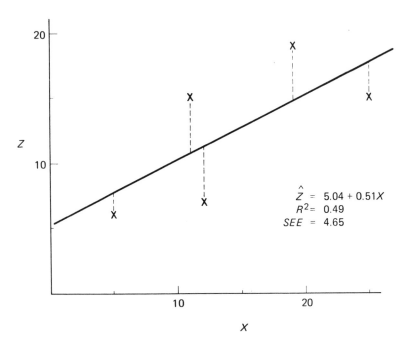

Figure 10.1
Predicting Z from X

Z. The vertical lines in the figure represent the residuals or prediction errors: the regression line underpredicts the value of Z for cases 2 and 4 and overpredicts for the others.

The question to ask now is "Is there something that distinguishes cases 2 and 4 from the other 3 cases?" If there is, we can use that information to adjust our predictions up or down from the first cut provided by the bivariate regression line. Our prediction rule might become something like, "Determine the value of X, then use as a first approximation $\hat{Z} = 5.04 + 0.51X$, and then if the case is like cases 2 or 4, add a little bit to the first approximation, otherwise, subtract a little bit." A look at table 10.1

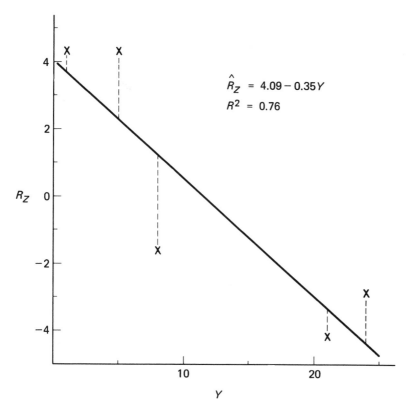

$$\hat{R}_Z = 4.09 - 0.35Y$$
$$R^2 = 0.76$$

Figure 10.2
Predicting residual Z (R_Z) from Y

shows that what distinguishes cases 2 and 4 are small values of Y relative to the other cases. Accordingly, we might want to plot a scattergram of the residual value of Z (which we will denote by R_Z) against Y, as in figure 10.2.

We see in figure 10.2 that we can come rather close to estimating the correction factor needed to improve our first approximation, since the error in prediction R_Z is itself rather well approximated by the regression line $\hat{R}_Z = 4.09 - 0.35Y$. We now have a fully quantitative form for our prediction rule: "Make a first approximation $\hat{Z} = 5.04 + 0.51X$, then add to that a correction factor $\hat{R}_Z = 4.09 - 0.35Y$." We can express the results of this two-stage prediction process as follows:

$$\hat{Z} = 5.04 + 0.51X$$
$$\hat{R}_Z = 4.09 - 0.35Y$$
$$\hat{\hat{Z}} = \hat{Z} + \hat{R}_Z$$
$$= (5.04 + 0.51X) + (4.09 - 0.35Y)$$
$$= 9.13 + 0.51X - 0.35Y,$$

where $\hat{\hat{Z}}$ represents the second cut at estimating Z (if planners can occasionally wear two hats—visionary and implementer—then so can their variables). Note that the prediction rule for $\hat{\hat{Z}}$ is a multivariate linear equation.

How well does this multivariate rule perform? We can compute the new residuals $(Z - \hat{\hat{Z}})$ and compare them to the old $(Z - \hat{Z})$ and to those arising from simply using the mean of Z as a prediction $(Z - \bar{Z})$. Using $\bar{Z}$ as the predictor, the sample standard deviation $S_Z = 5.64$ represent the typical error, whereas using $\hat{Z}$ gives $SEE = 4.65$, and using $\hat{\hat{Z}}$ gives $SEE = 2.78$; each stage of this regression analysis introduces major improvements in predictive power. The multivariate prediction rule leads to a rather substantial coefficient of determination: $R^2 = 0.88$. In practice, of course, we might still question the value of this analysis, since replacing the original set of only 5 values of Z with 3 regression constants (one intercept and two slopes) is hardly a great increase in portability. The real value of the example is in exposing the logic behind successive analyses of residuals and the improvements in predictive power that ensue.

The treatment of residuals just illustrated is neither the most used nor most powerful approach, although it is simple and started us off well. In practice we would not use Y itself to make the adjustment to the first cut but would instead use a purified version of Y. The reason is that X and Y

are correlated, if only slightly, so that some of the information encoded in the value of Y is already available in the value of X. For our second cut at approximating the value of Z, we want to use not the redundant information in Y but the fresh information. Therefore we purify Y by removing that part of the value of Y that is predictable from the value of X for the same case. It is then the residual value of Y which represents fresh information that (we hope) is predictive of the residual value of Z. Just as we would never bother to use $2X$ or $3X$ as a second predictor if we had already used X, since they carry no new information, neither should we expect that a set of Y values almost perfectly predictable from X will bring much new information to improve our predictions. That part of Y that cannot be

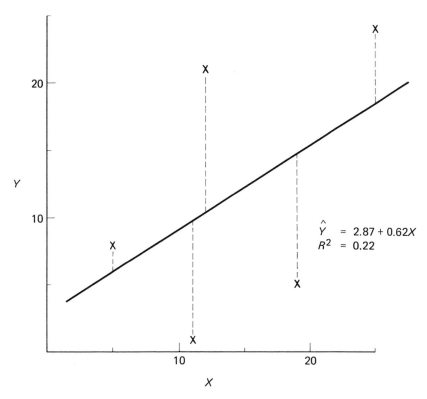

Figure 10.3
Predicting Y from X

predicted from X (the residual, R_Y) is the real news and the most useful predictor.

To examine the degree of redundancy between X and Y, we turn to the scattergram in figure 10.3. There is a positive association between the two attributes; although the association is weak, we will nevertheless improve our predictions by switching our choice of predictor variable from Y to $R_Y = Y - \hat{Y}$. Note that cases 2 and 4, which we previously noted as having low values of Y, are now the only cases with negative residuals R_Y, so their qualitative distinction remains; our hope is that the residuals R_Y will be better correlated with the residuals R_Z than was the original data Y.

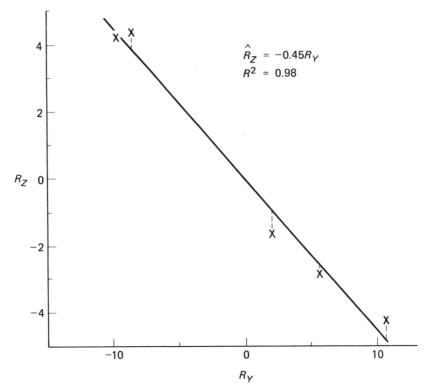

$$\hat{R}_Z = -0.45 R_Y$$
$$R^2 = 0.98$$

Figure 10.4
Predicting residual Z from residual Y (R_Y)

To test this we plot a scattergram of R_Z and R_Y, as in figure 10.4. Note that the correction factor R_Z needed to improve the first prediction made using X alone is given almost exactly by $\hat{R}_Z = -0.45R_Y$. We can calculate the correction factor almost perfectly and therefore produce a second approximation of very high quality. The steps are as follows:

$$\hat{Z} = 5.04 + 0.51X$$
$$R_Z = Z - \hat{Z}$$
$$\hat{Y} = 2.87 + 0.62X$$
$$R_Y = Y - \hat{Y}$$
$$\hat{R}_Z = -0.45R_Y$$
$$\hat{\hat{Z}} = \hat{Z} + \hat{R}_Z$$
$$= (5.04 + 0.51X) + [-0.45(Y - (2.87 + 0.62X))]$$
$$= 6.33 + 0.79X - 0.45Y.$$

This new multivariate regression equation performs even better than the last ($R^2 = 0.99$, $SEE = 0.75$); in fact, we could hardly expect more.

We can summarize this process more generally as follows, letting a, b, c, d, and e represent constants determined by least-squares analysis. First, use X to make a prediction of Z, and calculate the residuals R_Z for further analysis:

$$\hat{Z} = a + bX$$
$$R_Z = Z - a - bX.$$

Second, extract the fresh information in Y by predicting Y from X and calculating the residuals for use as the next predictors:

$$\hat{Y} = c + dX$$
$$R_Y = Y - c - dX.$$

Third, use the residuals R_Y to approximate the correction factors R_Z needed to improve the first prediction:

$$\hat{R}_Z = eR_Y.$$

Fourth, form the final prediction by adding the estimated correction factor to the first prediction:

$$\hat{\hat{Z}} = \hat{Z} + \hat{R}_Z$$
$$= (a + bX) + eR_Y$$

$$= (a + bX) + e(Y - c - dX)$$
$$= (a - ec) + (b - ed)X + eY.$$

Note that introducing Y into the analysis changes the coefficient of X from b to $b - ed$; this may even mean a change in the sign of the coefficient. Only when X and Y are not associated ($d \approx 0$) and / or the residuals R_Y are not associated with the errors in prediction R_Z ($e \approx 0$) does the introduction of Y leave the coefficient of X unchanged. The ideal is to have a small value of d (indicating that nearly all the information is Y is nonredundant) and a large value of e (indicating that the new information contained in the residuals R_Y is useful for determining the corrections to the first-cut predictions). Incidently, the roles of X and Y can be reversed in this process without changing the final outcome.

This same process could be applied to the introduction of a third predictor variable, say W. First, X and Y would be used to predict Z and a set of Z residuals calculated. Next, X and Y would be used to predict W in the same way and a set of W residuals determined. Then the W residuals would be used to predict the Z residuals. Finally, working back algebraically would eventually yield a single linear equation predicting Z from W, X, and Y. In practice, such a sequence of analyses of residuals would be time consuming and error-prone if done by hand. In fact, computers are almost always used for multivariate regression analysis, and even computers use a different computational scheme. Nevertheless the sequential method just outlined was stressed in part because it will permit you to do small three-variable problems with just a pocket calculator if you have to, but mostly so that you could see the logic of multivariate regression. The key is whittling down the size of residuals in the predicted variable by stages, at each new stage introducing predictors that have been purged of their linear associations with the predictor variables of previous stages: residuals predicting residuals.

Example: Predicting Municipal Expenditures from Population Size and Education

In planning practice we are rarely able to obtain a coefficient of determination as high as $R^2 = 0.99$ in any circumstances, let alone when using only two predictor variables. Neither are we often so fortunate as to work with predictor variables that are only weakly correlated among themselves. A more realistic example follows.

Suppose we want to develop predictions of the per capita expenditures (*EXP*) of Massachusetts cities and towns. There are many ways to try to predict *EXP*: from the size of the community, from its political party preferences, from its demographic makeup, and so forth. We will try to predict *EXP* from the community's income level (*INC*) as measured by median family income. Do towns whose residents have higher incomes spend more or less per capita than towns with lower-income residents? If I told you that a certain town had a median family income $2,000 greater than that of a neighboring town, which town would you expect to have greater expenditures per capita?

You probably suspect, quite rightly, that variables besides income will matter. Since we are using attributes of the population itself to assist our predictions, let us consider another: median years of education (*ED*). Perhaps citizens with a good deal of education come to value education highly, create pressure for high levels of school expenditures, and thereby drive up total expenditures per capita. Education level is a reasonable choice for a second predictor variable.

The multiple regression model predicting expenditures per capita from both median family income and median years of school completed will have the form

$$\widehat{EXP} = B_0 + B_1 ED + B_2 INC.$$

Our prediction of any community's value of *EXP* would begin at the level B_0, then add B_1 dollars for every year of median education (or subtract, if the sign of B_1 is negative), then add B_2 dollars for every dollar of median family income. We might hope that the coefficient B_2 would isolate the impact of income and B_1 isolate the impact of education. Of course, we would expect income and education to be associated themselves (multicollinearity), and successful interpretation of the effects of income and education will depend on the presence of mixes of the two variables in the data base. We will return to these issues of reading the regression coefficients later. For now our concern is prediction, not interpretation.

Let us review how to use the information about education to derive the multivariate regression estimate. In the bivariate case we would use income to predict expenditures. In the three-variable case we first use education to predict both expenditures and income. The residuals in

Table 10.2
Data for 20 Massachusetts communities relating education and income to municipal
expenditures per capita

	1970 median years of school	1970 median family income	1977 municipal expenditures (in thousands of dollars)	1975 population	Expenditures per capita
1. Easthampton	11.4	10,599	7,720	15,084	512
2. Oxford	11.8	10,621	5,823	10,822	540
3. Lawrence	10.4	9,507	33,717	67,515	499
4. Lexington	13.5	17,558	27,573	32,477	849
5. Marlborough	12.3	11,415	19,825	30,249	655
6. Medfield	12.8	15,609	7,026	10,031	700
7. Weymouth	12.4	11,631	37,612	56,854	662
8. Acushnet	10.2	9,691	4,101	8,439	486
9. Swansea	11.4	10,277	7,339	15,052	488
10. Salem	12.0	9,861	35,819	38,545	929
11. Wakefield	12.4	12,412	17,722	26,041	681
12. Provincetown	12.0	7,146	4,201	3,947	1,064
13. Hull	12.4	10,677	8,645	10,572	818
14. Wrentham	12.1	12,382	3,949	7,342	538
15. Weston	15.2	23,430	11,125	11,478	969
16. Lincoln	14.9	17,361	4,775	6,374	749
17. Chelsea	11.1	8,973	18,249	25,066	728
18. Blackstone	10.4	9,494	3,222	6,486	497
19. Palmer	11.3	10,203	5,581	11,755	475
20. South Hadley	12.2	11,091	7,367	16,568	445
Sample mean	12.1	12,002	13,471	20,535	664
Sample standard deviation	1.3	3,799	11,387	17,275	185

the prediction of expenditure represent the unexplained portion of expenditures, the part of the expenditure not predicted by education alone. The residuals in the prediction of income represent the fresh information in the income data, the amount of income above or below that which we would expect based on educational achievement. We will use the income residuals to predict the expenditure residuals, then work back algebraically to derive the multivariate regression equation in its usual form.

To illustrate the calculations in detail, I selected 20 Massachusetts communities for analysis. Data for these cities are given in table 10.2. Note that the education and income data, the population data and the expenditure data are for different years. The analysis would be more satisfactory if all data were contemporaneous, since the values of the predicting variables will shift to some extent over time; however, it is

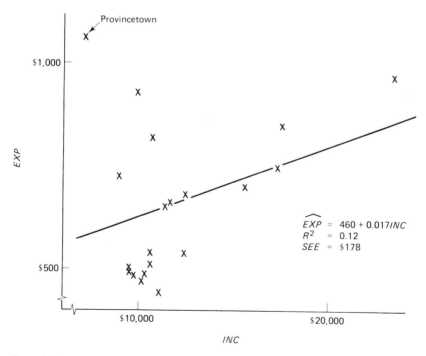

Figure 10.5
Predicting municipal expenditures per capita (*EXP*) from median family income (*INC*)

not uncommon to be stuck with rather dated data. Not also that table
10.2 opens the possibility of using population size as an additional
predicting variable. Although we will not do so, we could extend our
approach to incorporate population as a fourth variable.

The simple prediction of per capita municipal expenditure from median
family income is illustrated in figure 10.5; the bivariate regression equation
is

$$\widehat{EXP} = 460 + 0.017INC$$
$$R^2 = 0.12$$
$$SEE = \$178.$$

There seems to be a slight positive association between income and
expenditure, although the regression line provides a poor fit to the data
points. The regression slope, which increases the estimate of expenditure

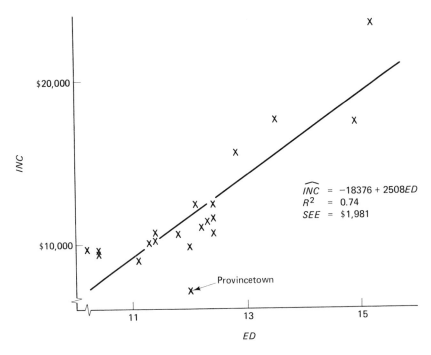

Figure 10.6
Predicting median family income from median years of school (*ED*)

by $17 for every $1,000 of income, would be greater were it not for the outlying point (Provincetown) in the upper left of the scattergram. Provincetown's special nature—a seasonal resort—might warrant its exclusion from analysis for some purposes, but we will continue to include it for this example. The general impression from figure 10.5 is that the greater the income of a community's residents, the more they spend for municipal services. Will this relationship persist after we introduce education into the analysis?

The degree of redundancy between income and education is evident in figure 10.6. The strong positive association clusters the data rather tightly around the regression line

$$\widehat{INC} = -18376 + 2508ED$$
$$R^2 = 0.74$$
$$SEE = \$1,981.$$

There are few communities with low education and high income, or vice versa, so the extent to which we can resolve the separate impacts of income and education will be somewhat limited. (Note that, since we expect few or no towns with median income below about $5,000, a curve would doubtless give a better fit to the data than the straight line shown in the figure.) Provincetown continues to be an outlier, with median income well below what we would expect for a community with median education of 12 years.

The utility of using education to predict expenditures is evident in figure 10.7. The fit of the regression line

$$\widehat{EXP} = -341 + 83ED$$
$$R^2 = 0.34$$
$$SEE = \$154$$

is moderate but still much better than that using income to predict expenditures. The greater the educational level of the citizens, the greater the expenditure per capita. At issue is how much of the discrepancy between the expenditure levels and the regression line—the residuals—can be eliminated by taking into account both the income and the education of the community's residents.

The regression residuals are listed in table 10.3. The first column presents the errors in predicting expenditures from income (see figure

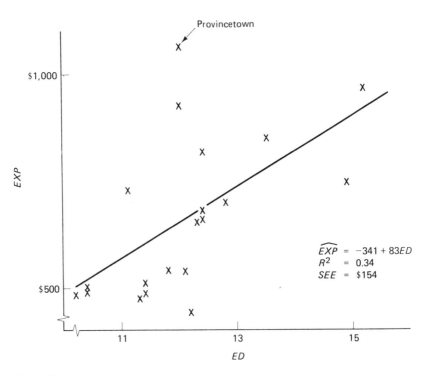

Figure 10.7
Predicting municipal expenditures per capita (*EXP*) from median years of school (*ED*)

10.5). The second column presents the errors in predicting expenditures from education (see figure 10.7); for 15 of the 20 towns the errors are smaller in absolute value when basing predictions on education rather than income. The third column presents the errors in predicting income from education (see figure 10.6); in contrast to the other columns, we would like these residuals to be large, since large residuals mean poor prediction of income from education, which in turn means that using the income data brings much fresh information to the problem. The last column in table 10.3 presents the errors made when using both income and education to predict expenditures, by using the residuals in the third column to predict the residuals in the second column. (Be 'sure you understand this last point.)

A comparison of the second and fourth columns is informative. The

Table 10.3
Regression residuals

	(1)	(2)	(3)	(4)
	Residual EXP after accounting for INC	Residual EXP after accounting for ED	Residual INC after accounting for ED	Residual EXP after accounting for ED and INC
1. Easthampton	$\$-128.37$	$\$\ -93.25$	$\$\quad 378.11$	$\$\ -82.13$
2. Oxford	-100.75	-98.46	-603.28	-116.20
3. Lawrence	-122.83	-23.23	$1,794.58$	29.55
4. Lexington	90.44	69.40	$2,069.33$	130.26
5. Marlborough	0.77	-24.97	$-1,063.51$	-56.25
6. Medfield	-25.46	-21.49	$1,876.26$	33.69
7. Weymouth	4.10	26.28	$-1,098.36$	-58.58
8. Acushnet	-138.95	-19.63	$2,480.27$	53.31
9. Swansea	-146.90	-117.25	56.11	-115.60
10. Salem	301.16	273.93	$-1,864.97$	219.08
11. Wakefield	9.83	-7.28	-317.36	-16.61
12. Provincetown	482.27	408.93	$-4,579.97$	274.24
13. Hull	176.30	129.72	$-2,052.36$	69.36
14. Wrentham	-132.66	-125.37	405.18	-113.45
15. Weston	109.01	48.26	$3,776.94$	159.34
16. Lincoln	-6.22	-146.83	$-1,639.52$	-195.05
17. Chelsea	115.24	147.65	-495.35	133.08
18. Blackstone	-124.61	-25.23	$1,781.58$	27.16
19. Palmer	-158.65	-121.95	232.96	-115.10
20. South Hadley	-203.73	-226.67	$-1,136.66$	-260.10
Sum of residuals[a]	-0.01	0.00	-0.02	0.00
Sum of squared residuals[b]	$570,913$	$427,090$	$70,672,520$	$365,968$

[a] Would equal zero except for rounding errors.
[b] Used in computing R^2 and SEE.

prediction errors made using both education and income are larger in absolute value in 12 of the 20 cases than those when using education alone. In this sense the multivariate regression is inferior to the bivariate regression. However, the total sum of squared residuals is less in the multivariate case, and this sum is the criterion we chose for curve fitting. Most of the improvement can be traced to just one community—Provincetown—whose large unexplained expenditure ($408.93) is reduced substantially (to $274.24) when income is added to the analysis as a predictor variable. The least-squares criterion gives extra attention to such deviant cases, greasing squeaky wheels at the expense of the other data cases.

To show how the final form of the multivariate regression equation comes about, step backwards through the sequence of bivariate regressions. The regression of residual expenditure (column 2 of table 10.3) on residual income (column 3) yields

$$\text{Residual } \widehat{EXP} = -0.029 \text{ Residual } INC.$$

Now since

$$\text{Residual } INC = INC - (-18376 + 2508ED),$$

it follows that

$$\text{Residual } \widehat{EXP} = -0.029[INC - (-18376 + 2508ED)].$$

Further, since

$$\widehat{EXP} = \widehat{EXP} + \text{Residual } \widehat{EXP},$$

and

$$\widehat{EXP} = -341 + 83ED,$$

we have

$$\widehat{EXP} = \{-341 + 83ED\} + \{-0.029[INC - (-18376 + 2508ED)]\}$$

which, upon rearranging, gives the formula

$$\widehat{EXP} = -874 + 156ED - 0.029INC.$$

Actually, keeping more decimal places in the calculations leads to

$$\widehat{EXP} = -882 + 157ED - 0.029INC.$$

Note that the sign of the coefficient of *INC* is negative in the multivariate equation but positive in the bivariate equation.

The summary statistics for the multivariate regression are

$$R^2 = 1 - \frac{\sum\limits_{i=1}^{20} [EXP_i - (-882 + 157ED_i - 0.029INC_i)]^2}{\sum\limits_{i=1}^{20} (EXP_i - 664)^2} = 0.44$$

$$SEE = \sqrt{\frac{1}{20 - 2 - 1} \sum\limits_{i=1}^{20} [EXP - (-882 + 157ED - 0.029INC)]^2}$$

$$= \$147.$$

So using income and education to predict expenditures per capita typically gets us to within about \$147 of the actual value; this compares to getting within about \$185, using only the sample mean as the prediction, and \$154, using education alone. As noted in chapter 9, an even better test of prediction power would apply the regression equation to new data cases.

Dummy Variables

Sometimes we are naturally led to predicting variables that are nominal or ordinal, not metric. There is a way to include these in multiple regression analysis—we are not restricted to using just metric predicting variables. We will retain the restriction that the predicted variable be metric (alternatives to regression analysis can accommodate binary predicted variables; these alternatives include *logit* and *probit* analysis).

For instance, suppose we want to predict annual income from sex. We can define a *dummy variable SEX* which takes on the value 0 if the person is female and the value 1.0 if the person is male (the choice is arbitrary—we could as well assign the value 0 to males and 1.0 to females). Then we would compute the bivariate regression

$$\widehat{INCOME} = B_0 + B_1 SEX.$$

Suppose analysis of our data gave the estimates

$$\widehat{INCOME} = \$8{,}500 + \$3{,}175 SEX.$$

Then our prediction of a woman's income would be $\$8{,}500 + \$3{,}175 \times 0 = \$8{,}500$; our prediction for a man would be $\$8{,}500 + \$3{,}175 \times 1 = \$11{,}675$. The coefficient B_1 summarizes the sex differential in income.

Now suppose we felt that race as well as sex was important to the prediction of income. We might define two categories of race—white and nonwhite—and (arbitrarily) assign the value $RACE = 0$ to whites and $RACE = 1.0$ to nonwhites. Then we could determine the multiple regression equation

$$\widehat{INCOME} = B_0 + B_1 SEX + B_2 RACE.$$

We would expect the values of B_0 and B_1 to change from the bivariate case as we add the income effects of race to those of sex. Suppose analysis provided the estimates

$$\widehat{INCOME} = \$7{,}100 + \$2{,}982 SEX - \$4{,}158 RACE.$$

The predicted income will depend on both race and sex in the following way:

White male	$B_0 + B_1$	$= \$10{,}082$
White female	B_0	$= \$\ 7{,}100$
Nonwhite male	$B_0 + B_1 + B_2$	$= \$\ 5{,}924$
Nonwhite female	$B_0 + B_2$	$= \$\ 2{,}942.$

Note that this last model assumes that the race differential is the same for either sex and the sex differential the same for either race. A more general model would use a new dummy variable defined as the product of $RACE$ and SEX, which takes the value 1.0 for nonwhite males and 0 for all others. Now the prediction rule would be

$$\widehat{INCOME} = B_0 + B_1 SEX + B_2 RACE + B_3 RACE \times SEX,$$

and the difference in income between, for example, white and nonwhite males need not be the same as the difference between white and nonwhite females. If the estimates are

$$\widehat{INCOME} = \$6{,}950 + \$3{,}115 SEX - \$4{,}008 RACE - \$987(RACE \times SEX),$$

then the income predictions would be

White male	$B_0 + B_1 = \$10,065$
White female	$B_0 = \$ 6,950$
Nonwhite male	$B_0 + B_1 + B_2 + B_3 = \$ 5,070$
Nonwhite female	$B_0 + B_2 = \$ 2,942.$

(These same estimates could also have been gotten by computing mean income separately for each of the four categories of individuals.)

We can extend this result to include any categorized independent variable with more than just 2 categories. Suppose we wanted to predict *RENT* from knowledge of the physical appearance of the building containing the dwelling unit, and suppose further that "appearance" is an ordinal variable with 5 categories: very bad, poor, fair, good, and very good. We create 4 dummy variables (always one fewer than the number of categories) called X_1 through X_4 and give them values as follows:

Category of Appearance	X_1	X_2	X_3	X_4
Very bad	0	0	0	0
Poor	1	0	0	0
Fair	0	1	0	0
Good	0	0	1	0
Very good	0	0	0	1

Thus each dwelling unit in our data base would have associated with it 5 values: *RENT*, X_1, X_2, X_3, and X_4. The regression equation would be

$$\widehat{RENT} = B_0 + B_1 X_1 + B_2 X_2 + B_3 X_3 + B_4 X_4.$$

The predicted difference in rent between, say, a unit in a building with "poor" appearance and a unit in a "very good" building would therefore be $B_4 - B_1$. We could have specified our income prediction model the same way:

Category of individual	X_1	X_2	X_3
White male	0	0	0
White female	1	0	0
Nonwhite male	0	1	0
Nonwhite female	0	0	1

in which case the prediction equation would have been

$$\widehat{INCOME} = B_0 + B_1 X_1 + B_2 X_2 + B_3 X_3.$$

It is common to make predictions from a mixture of metric and dummy variables. Suppose we wish to predict annual operating costs for nursing homes. We suspect that both the size of the nursing home (as measured by the number of beds) and its type (proprietary or nonprofit) will influence costs. We might define a dummy variable

$TYPE = 1$ for proprietary homes
$\quad\quad\quad = 0$ for nonprofit homes

and predict

$$\widehat{COST} = B_0 + B_1 BEDS + B_2 TYPE.$$

This model implies that we expect the type of home to impact on costs only by adding to (or subtracting from) the intercept of the cost curve, as shown in figure 10.8. More generally, we could allow for the possibility that both the intercept and the slope would differ between the two

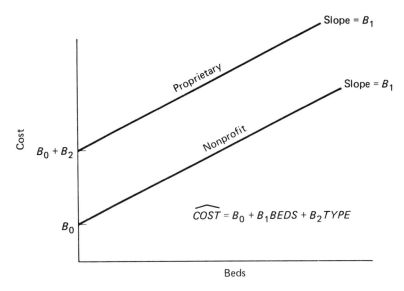

Figure 10.8
The use of a dummy variable to shift the intercept of a regression line

types of nursing homes (the marginal cost per bed might differ). This more general specification would include a new predicting variable formed by multiplying *BEDS* by *TYPE*:

$$\widehat{COST} = B_0 + B_1 BEDS + B_2 TYPE + B_3(BEDS \times TYPE).$$

The resulting prediction rules are depicted in figure 10.9, which shows a separate intercept and slope for each type of nursing home.

Transformations

The same types of transformations useful in bivariate regression are useful in multivariate regression as well. For instance, we might believe that a multiplicative (rather than additive) model specification would be useful in predicting the volume of rail trips between pairs of cities as a function of their combined populations (*POP*), the trip time (*TIME*), and the price of the trip (*PRICE*):

$$\widehat{TRIPS} = B_0 POP^{B_1} TIME^{B_2} PRICE^{B_3}.$$

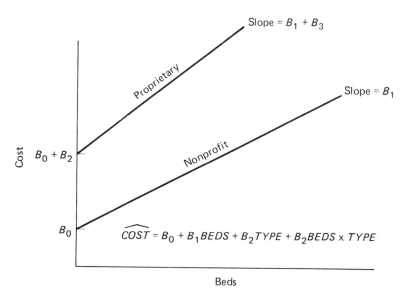

Figure 10.9
The use of a dummy variable to shift both the intercept and slope of a regression line

This specification (referred to in the economics literature as a "Cobb-Douglas type production function") can be converted to a linear additive model by taking logs:

$$\widehat{\log TRIPS} = \log B_0 + B_1 \log POP + B_2 \log TIME + B_3 \log PRICE.$$

The unknown constants B_1, B_2, and B_3 are elasticities and represent the predicted percentage change in *TRIPS* associated with a 1 percent change in *POP*, *TIME*, or *PRICE*, respectively. For instance, a 1 percent change in *PRICE* is expected to correspond to a B_3 percent change in *TRIPS*. (This model assumes constant elasticities; more complex models might assume, for example, that price elasticity will itself vary according to the level of the *PRICE*.) A multiplicitive model differs from an additive model in the following critical way. In a multiplicative model the predicted change in *TRIPS* associated with, say, a $10 increase in *PRICE* would depend on the values of *POP* and *TIME*; in a linear model the predicted change in *TRIPS* would be the same regardless of the values of *POP* and *TIME*. Multiplicative models of the form we gave in predicting rail trips can be thought of as the multivariate extension of the power curve fit introduced in chapter 9.

Another transformation that is often helpful when the predicted variable is a fraction (or proportion) is the *logit* transformation. When the predicted variable is a fraction, it can only take on values in the range 0 to 1.0, so we would like to avoid predictions falling outside this range. However, a model like

$$\widehat{FRACTION} = 0.83 + 0.12X + 0.34Y$$

would predict fractions greater than 1.0 or less than 0 for certain combinations of *X* and *Y* (for example, $X = 0.9$, $Y = 0.2$). If such combinations of *X* and *Y* are possible, the predictions will not be directly usable. However, if we could expand the range of the predicted variable from negative infinity to positive infinity, this problem would never arise. A transformation of the form

$$LOGIT = \log\left(\frac{FRACTION}{1.0 - FRACTION}\right)$$

will do the trick, as shown in figure 10.10. We can estimate coefficients in

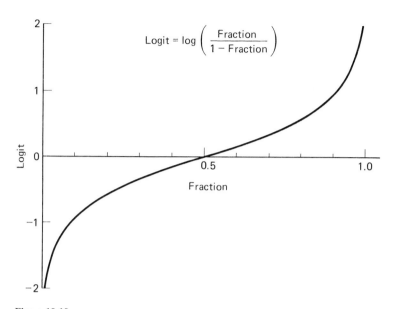

Figure 10.10
The logit transformation

$$\widehat{LOGIT} = B_0 + B_1 X + B_2 Y$$

and later obtain estimates of *FRACTION* by undoing the logit transformation

$$\widehat{LOGIT} = \log \frac{\widehat{FRACTION}}{1 - \widehat{FRACTION}}$$

$$10^{\widehat{LOGIT}} = \frac{\widehat{FRACTION}}{1 - FRACTION}$$

$$\widehat{FRACTION} = \frac{10^{\widehat{LOGIT}}}{1 + 10^{\widehat{LOGIT}}}.$$

(You should note that logit analysis produces a problem we will discuss called "heteroscedastic residuals.")

Testing Whether the Regression Equation Predicts Better than the Sample Mean

There is no doubt that multivariate and even bivariate regression is more costly and complex than simply using the sample mean as a predictor, just as analysis of contingency tables is more costly and complex than using the modal category of the unconditional sample distribution as a predictor. When we performed analysis of contingency tables, we found the descriptive level of significance useful in deciding whether associations apparent in the tables could plausibly be dismissed as artifacts of the sampling process, meaning that the attributes chosen to condition our predictions were really of no assistance. In the same way we can test whether the predictor variables in a regression analysis add any predictive power beyond that available in the sample mean of the predicted variable.

The general mechanism of significance testing is the same in regression analysis as in the analysis of contingency tables. We first form a *null hypothesis of independence*. In the case of regression analysis, this amounts to the hypothesis that none of the variables chosen as predictors is associated in the population with the predicted variable. In other words, the regression equation is hypothesized to do no better than the sample mean as a predictor. This comparison with the sample mean is the key to the second step in significance testing: choosing a *test statistic*.

We already have in the coefficient of determination R^2 a summary measure that compares the predictive power of the regression equation to that of the sample mean. Thus we might operationalize the null hypothesis of independence by stating it as the hypothesis that the true value of R^2 in the population is zero. Unfortunately, this straightforward choice of a test statistic does not lend itself to ease of analysis at the next step, so we choose instead a variant of R^2. If n is the number of cases in the sample and k the number of predictor variables, then the test statistic is the value of

$$F_{obs} = \frac{(R^2/k)}{(1 - R^2)/(n - k - 1)}.$$

When $R^2 = 0$, then $F_{obs} = 0$; as R^2 approaches 1.0, F_{obs} grows without limit.

The third step is to determine the *sampling distribution of the test statistic*. The distribution of F_{obs} has been well tabulated (see appendix D) when the null hypothesis of independence is true and the regression residuals have

1. a Gaussian distribution,
2. constant variance throughout the range of the predicted variable,
3. zero mean throughout the range of the predicted variable,
4. zero correlation between successive residuals.

(These four assumptions are necessary for any customary inference in regression analysis.)

Even when the predictor variables are completely independent of the predicted variable in the population, there will inevitably be some accidental association in the sample, so the value of R^2 will not be exactly 0, and therefore neither will the value of F_{obs}. Just as with χ^2_{obs},

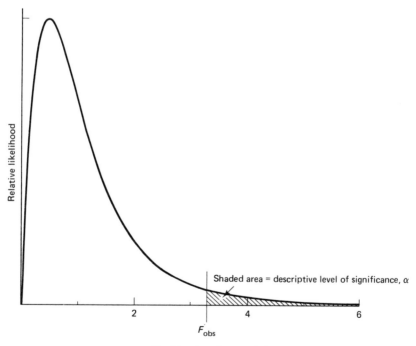

Shaded area = descriptive level of significance, α

Possible value of test statistic

Figure 10.11
The F distribution with (4, 16) degrees of freedom, showing the descriptive level of significance

small values of F_{obs} are to be expected due to accidents of the sampling process, but large values cast doubt on the null hypothesis. When the four conditions above are true, the test statistic F_{obs} has the *F distribution* with degrees of freedom k and $n - k - 1$. The quantity k is often referred to as the "numerator degrees of freedom" and $n - k - 1$ is called the "denominator degrees for freedom" because of their respective locations in the numerator and denominator of the formula for F_{obs}. The *F* distribution with $df = (4, 16)$, that is, with degrees of freedom 4 and 16, corresponding to $k = 4$ and $n = 21$, is shown in figure 10.11. We learn from figure 10.11 that when using $k = 4$ predictor variables to analyze a set of $n = 21$ cases, the value of F_{obs} will quite often reach a value in the vicinity of about 1.0 or 2.0 simply because of sampling accidents, whereas values of F_{obs} greater than, say, 3.0 are hard to explain as sampling accidents. Working backwards in the formula for F_{obs}, we find that a value of $F_{obs} = 3.0$ corresponds to a value of $R^2 = 0.43$. Therefore if we have data on just 21 cases and we use as many as 4 predictor variables, we should not be much impressed with values of R^2 below about 0.40, since these could easily arise accidently even when the null hypothesis is true. The descriptive level of significance of F_{obs} is defined like that for χ^2_{obs}:

α = descriptive level of significance

$$= \mathrm{Prob} \left[F_{df = k, \, n-k-1} \geq F_{obs} \left| \begin{array}{l} \text{null hypothesis of} \\ \text{independence is true} \end{array} \right. \right].$$

The value of α is obtained from the table of the *F* distribution in appendix D, either directly or by interpolation.

For instance, consider again the bivariate regression analysis which predicted full-value tax rate from population in a sample of $n = 11$ Massachusetts communities:

$$\widehat{TAX} = 21.66 + 0.72 POP$$
$$R^2 = 0.40.$$

For this case

$$F_{obs} = \frac{R^2 / k}{(1 - R^2)/(n - k - 1)}$$
$$= \frac{0.40/1}{(1 - 0.40)/(11 - 1 - 1)} = 6.00.$$

Now the descriptive level of significance is

$\alpha = \text{Prob}\,[F_{df=1,9} \geq 6.00\,|\,\text{null hypothesis is true}].$

Reference to the table in appendix D shows that

$\text{Prob}\,[F_{df=1,9} \geq 5.12] = 0.05,$

from which we conclude that $\alpha < 0.05$. Consequently, we place little stock in the argument that the value $R^2 = 0.40$ was generated by sampling accident. Instead we lean strongly toward the view that the regression rule will serve as a better predictor than will the sample mean.

As a second example, consider the analysis made earlier in this chapter of the use of a community's median family income and median years of schooling to predict per capita municipal expenditures:

$$\widehat{EXP} = -882 + 157ED - 0.029INC$$
$$R^2 = 0.44.$$

In this case

$$F_{obs} = \frac{(0.44/2)}{(1 - 0.44)/(20 - 2 - 1)} = 6.68.$$

Therefore

$\alpha = \text{Prob}\,[F_{df=2,17} \geq 6.68\,|\,\text{null hypothesis is true}].$

Now since the table of the F distribution reveals that

$\text{Prob}\,[F_{df=2,17} \geq 3.59\,|\,\text{null hypothesis is true}] = 0.05,$

it would appear safe to conclude that $\alpha < 0.05$, and the relative advantage of the regression analysis compared to the sample mean seems clear in this case as well (but see the discussion that follows about the violation of assumptions for inference in this example).

Checking the Assumptions of Inference in Regression Analysis

The use of the F distribution to establish the descriptive level of significance of F_{obs} (and therefore of R^2) depends on the validity of the four assumptions made about the regression residuals. Likewise the validity of the highest density regions for predictions (see chapter 9) depends on the validity of these assumptions. When these assumptions are not met, the customary inference procedures are invalid, often to an unknown extent. There are techniques available to cope with some of the violations of these assumptions (see Kmenta's book); we will have to be content with an informal discussion of when to entertain serious doubts about their validity.

The assumption that the residuals are Gaussian can be checked in the same way that we checked the distribution of individual sample values when attempting to construct a highest density region for the sample mean (chapter 7). The graphical technique of ordering the values from smallest to largest, assigning the ith case the value $\dfrac{i}{n+1}$, then plotting on probability paper, should serve well when the number of cases is reasonably small. Some computer packages will plot standardized residuals, which should be eyeballed for approximate symmetry and for an appropriate thinning out with size. Glaring deviations from the Gaussian shape should warn you away from constructing highest density regions or determining the descriptive level of significance; otherwise proceed with the inference procedures as reasonable approximations. If you have only a few data points you will not be able to do much to verify the Gaussian assumption, but neither will you be likely to disprove it; in any event your real problem in such cases will probably be getting more data, not improving the estimates of uncertainty.

The assumption that the residuals have a constant variance can be checked informally by looking at a plot of the residuals $Y - \hat{Y}$ against the predicted values $\hat{Y}$ (or at a plot of the standardized values of both). If there is a marked pattern of greater spread in some portions of the scattergram than in others, then *heteroscedasticity* or nonconstant variance is a problem. Otherwise the residuals are said to be *homoscedastic*, and inference can proceed unabashed (provided the other assumptions are tenable). Be aware, however, of a kind of trick that can be played on the eye by a clustering of cases which have nearly the same predicted value. If many cases are bunched together, there is a good chance that at least one will have a very high or very low residual just because the distribution is sampled more often, even when the residuals are truly homoscedastic (for example, samples of 5 have a range that averages twice as wide as samples of 2 from the same Gaussian distribution). This bunching can create a false impression of heteroscedasticity.

The assumption that the residuals have zero mean throughout the range of the predicted variable can be tested informally with the same type of residual plot as is used to test for heteroscedasticity: a plot of $Y - \hat{Y}$ on the vertical axis and $\hat{Y}$ on the horizontal. If, for instance, all the underestimates (positive residuals) are paired with small values of $\hat{Y}$ and all the overestimates (negative residuals) with large values of $\hat{Y}$, then there is systematic variability in the residuals that might be exploited to improve predictions.

The assumption that the residuals have constant variance is most often problematical when the data are *cross-sectional*, deriving from a snapshot of many cases at one instant of time. When the data are *time series*, arising from repeated observations of a single case over time, then the problem of heteroscedasticity usually gives way to the problem of *autocorrelation*, in which successive residuals are not independent. This means that knowledge of the residual for one time period can be used to predict the size of the residual at a subsequent time period (usually the next). One graphical way to detect this problem is to plot a scattergram of the residual for one time period against the residual for a subsequent time: if a strong association is apparent, autocorrelation is a problem. Another common approach is to compute the *Durbin-Watson statistic*

$$d = \frac{\sum\limits_{t=2}^{n} (r_t - r_{t-1})^2}{\sum\limits_{t=1}^{n} r_t^2},$$

where r_t is the residual for the tth time period and r_{t-1} is the preceding residual. Suppose the residual at one time is highly predictive of the next residual in the sense that $r_t \approx r_{t-1}$ (there is a high positive correlation between successive residuals). Then each term in the numerator of the expression for the Durban-Watson statistic d will be very small, and the value of d will turn out to be just a little bigger than 0. Thus a small value of d is a sign that there is a strong positive correlation between successive residuals. On the other hand, if successive residuals have a high negative correlation in the sense that $r_t \approx -r_{t-1}$, then each numerator term $(r_t - r_{t-1})^2 \approx (2r_t)^2$, so the value of d will be near 4.0 in this case. The general conclusion is that a middle-sized value of d near 2.0 indicates smooth sailing, whereas values of d closer to 0 or 4.0 suggest that

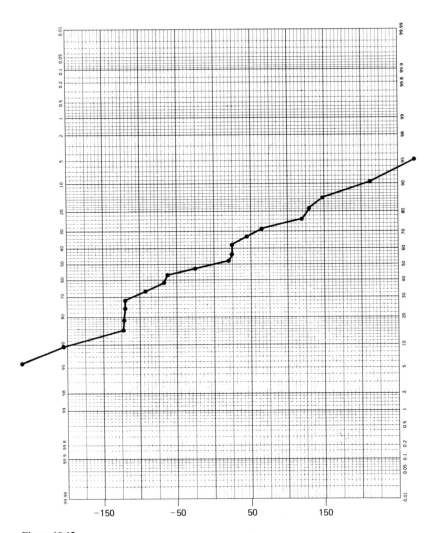

Figure 10.12
Plot of regression residuals on probability paper to test assumption of Gaussian distribution

autocorrelation is present to a major degree (Kmenta's book offers a formal test for the presence of autocorrelation based on the Durbin-Watson statistic).

Incidently, autocorrelation of residuals can be a problem with cross-sectional data too if the data have a spatial character. For instance, predicting snow-removal expenses in a community from its total street-miles may lead to a situation in which an under- or overestimate in one community is likely to be predictive of the residual expense in adjacent communities, which presumably share much the same weather. Since planners deal so often with spatially oriented data, you should be alert to the possibility of correlated residuals even in cross-sectional data.

To see the application of these techniques, consider the residuals of the regression predicting per capita municipal expenditures from median years of school completed and median family income:

$$\widehat{EXP} = -882 + 157ED - 0.029INC.$$

Since in this case we have neither time-series data nor any substantive reason to expect spatial correlation, we will skip the analysis for correlated residuals. To test the assumption that the residuals have a Gaussian distribution, we plot them on probability paper, as in figure 10.12. The figure reveals that the points lie tolerably close to a straight line, so the Gaussian assumption seems safe enough. To test the assumption of

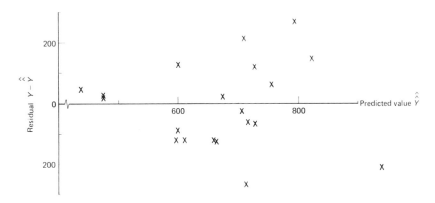

Figure 10.13
Plot of regression residuals against predicted values to test assumptions of zero mean and constant variance of residuals

homoscedasticity, we plot the residuals against the predicted values, as in figure 10.13. That figure reveals a pattern suggestive of heteroscedasticity, with greater dispersion of the residuals for higher predicted expenditures. On reflection, this heteroscedasticity should not be unexpected —the same sort of behavior is observed when predicting personal expenditures from income. One plausible explanation is that, when a community has a low level of expenditure, there is relatively little leeway for discretionary spending, most monies being devoted to basics. On the other hand, when the level of spending is potentially greater, there is more room for discretion and hence more variability around the normal level of expenditures. The assumption that the residuals have zero mean throughout the range of the predicted variable can also be checked using figure 10.13. While the figure shows a trend in the dispersion of the residuals, it shows no trend in their central tendency, so the only violation of the assumptions for inference involves heteroscedasticity.

Now the existence of heteroscedasticity does not in itself mean that the regression equation is not a useful prediction tool. In fact, the least-squares line can be useful even if all four assumptions are violated. However, it may not be proper to proceed with the customary inference procedures if any one of the four assumptions is violated. In serious work you would perform formal statistical tests of the validity of the assumptions and if necessary employ various data weightings or transformations to restore their validity.

Regression and Causality

To this point we have been treating regression analysis solely as a tool for prediction. However, for better or for worse, regression analysis—like analysis of contingency tables—is often used to test or to create causal theories, so we must consider this use as well. In all fundamental respects the remarks on causality and contingency tables in chapter 8 apply to regression analysis as well.

The most direct way to see the impact on one variable of a change in a second variable is in fact to change the second. However, opportunities for such direct control are not plentiful, so regression analysis of observational data is a very common activity. In such studies the attention of the analyst centers not so much on prediction but on the values of the regression coefficients.

It is tempting to read a *regression coefficient* as "the amount of change in the dependent variable that would be produced by a change of one unit in the independent variable, with all other independent variables held constant." If this interpretation were really true, planners could know with great confidence ahead of time what would happen if they were to initiate certain changes in the systems for which they plan. Unfortunately, although this interpretation is usually made, it is almost always improper in a planning context. There are at least three reasons why this interpretation is likely to be misleading.

First, we may not really be interested in the answer to the question as it is commonly phrased. While it may be theoretically interesting to know what would happen if, say, X_1 were varied while X_2 was held constant, it may be that X_2 is itself related to X_1, so that changing X_1 would entrain changes in X_2 as well, leading to rather different net effects than expected. This would be the case, for instance, if there were multivariate causation of the form

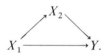

Here holding X_2 constant would keep us from knowing what would really happen to Y if we manipulated X_1. Of course, if X_1 and Y were linked solely through X_2 as a causal intermediary

$$X_1 \longrightarrow X_2 \longrightarrow Y,$$

then it is clear that varying X_1 but somehow compensating to keep X_2 constant would thwart any change in Y.

Second, the association summarized in a regression analysis may not be causal after all. For instance, if there is a spurious correlation between X_1 and Y because both are caused by some variable X_3 not in the analysis

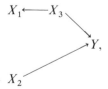

then manipulating X_1 will not change the value of Y.

Third, the value of the regression coefficient may not be useful even if there are simple, direct causal links of the form

One reason for this is multicollinearity, which manifests itself in regression analysis by producing very uncertain estimates of the regression coefficients. A second reason is *error in measurement* of the variables. Whereas multicollinearity leads to very uncertain estimates, which nevertheless tend to be centered on the correct values, errors in measurement lead to coefficients that center on incorrect values. A third reason is the sensitivity of the least-squares criterion to outliers—either unrepresentative cases or outright blunders in the data. Each of these problems will be discussed in more detail. All three are important for practice, since some degree of multicollinearity and error in measurement are present in just about any regression analysis undertaken by planners, and key-punching and other processing mistakes are all too common. It is quite prudent to maintain a healthy skepticism about the numerical values of regression coefficients.

Every planner would like to know the answer to questions like, "What will happen to Y if I decrease X_3 by one unit?" Under ideal circumstances, regression analysis could help with such questions. Unfortunately, under realistic circumstances, it may be very bad practice to regard the regression coefficient of X_3 as a safe and useful answer. Do not expect to do well by blindly grinding out a single regression analysis of the first available data set. Instead, develop well-thought-out theories, seek out data of high quality, and replicate your analysis on as many data sets as possible. Regression analysis is one of the best tools we have for testing or building causal theories, but you cannot surrender too much of your responsibility for understanding urban systems to a curve-fitting procedure.

Multicollinearity

The problem of multicollinearity was first treated in chapter 8. This discussion will focus on the reliability of estimates of regression coefficients. Regression coefficients are random variables: the coefficients estimated from one data set will always differ to some extent from those estimated from a different data set. In the same way that the sample mean is a random variable which fluctuates from sample to sample and therefore has a standard error, so each regression coefficient has a standard error.

Just as we used highest density regions centered on the sample mean to determine an interval with a specified probability of bracketing the true value of the coefficient in the population as a whole, so we can construct HDR's centered on a sample regression coefficient and having a specified probability of bracketing the true value of the coefficient in the population. When working with the sample mean $\bar{Y}$, the HDR had the form

$$HDR \text{ for population mean} = \bar{Y} \pm C_{n-1} \times SE_{\bar{Y}},$$

where

$SE_{\bar{Y}} = S_Y / \sqrt{n}$ is the standard error of the sample mean,
S_Y = sample standard deviation of Y,
C_{n-1} = a specified percentile of Student's t distribution with $n - 1$ degrees of freedom,
n = number of data cases.

Likewise when working with the regression coefficient of the predictor variable X_i, the HDR has the form

$$HDR \text{ for population coefficient} = B_i \pm C_{n-k-1} \times SE_{B_i},$$

where

B_i = estimated regression coefficient,
$SE_{B_i} = \dfrac{(SEE / S_{X_i})}{\sqrt{(n - 1)(1 - R^2_{X_i})}} = $ standard error of regression coefficient,
SEE = standard error of estimate of regression equation,
S_{X_i} = sample standard deviation of predictor variable X_i.

n = number of data cases,

k = number of predictor variables,

$R^2_{X_i}$ = coefficient of determination of the regression which predicts the value of X_i from all the other predictor variables,

C_{n-k-1} = a specified percentile of Student's t distribution with $n - k - 1$ degrees of freedom.

The *HDR* will be narrow, and therefore our estimate of the true value of the regression coefficient in the population will be firm, whenever the standard error is small. The standard error, in turn, will be small if

1. the regression predicts well, so that *SEE* is small,
2. the data set embodies wide-ranging experience with X_i, so that S_{X_i} is large,
3. the number of cases n is large relative to the number of predictor variables k,
4. the variable X_i adds fresh information in the sense that its value cannot be estimated well from the other predictors, as indicated by a low value of $R^2_{X_i}$.

The last observation is a restatement of a point first made when considering contingency tables: if predictors are too highly associated with each other, it will be impossible to isolate their individual effects. In the hypothetical redlining example in table 8.9, racial and economic attributes were so strongly associated that banks' failure to lend could not be attributed to one or the other factor. In the context of regression analysis, if the information available in variable X_i is so redundant that we can accurately forecast the value of X_i from knowledge of the other predictors, then we cannot expect a firm estimate of its coefficient. Multicollinearity makes it difficult to divide up the responsibility for Y.

For an illustration, return to the example of predicting municipal expenditures from income and education. The regression equation was

$$\widehat{EXP} = -882 + 157ED - 0.029INC$$
$$R^2 = 0.44$$
$$SEE = \$147.$$

In the sample of $n = 20$ communities there was a strong association between education and income ($R^2 = 0.74$; see figure 10.6); this multicollinearity will tend to muddy the waters. Table 10.2 provides the sample standard deviations: 1.3 years for education and \$3,799 for income. We have now assembled all the facts needed to compute the standard errors of the regression coefficients. Using the formula

$$SE_{B_i} = \frac{(SEE/S_{X_i})}{\sqrt{(n-1)(1-R^2_{X_i})}},$$

we have

$$SE_{ED} = \frac{(147/1.3)}{\sqrt{(20-1)(1-0.74)}} = 51$$

and

$$SE_{INC} = \frac{(147/3799)}{\sqrt{(20-1)(1-0.74)}} = 0.017.$$

The standard error is about one-third the size of the coefficient in the case of ED and is more than half the size of the coefficient of INC. To construct 95 percent HDR's, we note in appendix C that for $n - k - 1 = 20 - 2 - 1 = 17$ degrees of freedom the 97.5 percent point of Student's t distribution occurs at $C_{17} = 2.11$. Using the formula

$$HDR = B \pm C_{n-k-1} \times SE_B,$$

we find

95 percent HDR for $B_{ED} = 157 \pm (2.11) \times (51)$
$$= 49 \text{ to } 265,$$

while

95 percent HDR for $B_{INC} = 0.029 \pm (2.11) \times (0.017)$
$$= -0.065 \text{ to } 0.007.$$

(Of course, these are only approximate HDR's, since the earlier analysis of residuals discovered heteroscedasticity.) Both magnitudes are rather uncertain, and even the sign of the coefficient of INC is slightly in doubt. We could improve matters by adding more communities to the data base, especially communities with income levels not only far from the current sample mean of \$12,002 but also far from the values predicted from their education levels by

$$\widehat{INC} = -18376 + 2508ED.$$

Errors in Measurement

Very often planners use predictor variables that are numerical stand-ins for more fundamental but less measurable variables. Examples would be the use of education as a measure of socioeconomic status or the use of the age of a housing stock as a measure of its quality. Sometimes planners work with quantities that are precisely measurable in principle but poorly measured in practice, such as a family's money income when self-reported in broad categories during an interview designed to determine eligibility for services (in this case fallible memories, incentives to shade the truth, and aggregation of the data into categories all lead to deviations from the true figures). Not infrequently, additional errors are introduced into data sets during the process of preparing them for analysis. All of these sources of imperfection make themselves felt by biasing the values of the regression coefficients computed from the data, and gathering more data does not necessarily offer relief. Estimating the size of the bias is not easy, but Cochran has succeeded in analyzing the case of bivariate regression.

Let X and Y denote the true but unknown values, and let x and y denote their measured values. If h and d represent the errors of measurement of X and Y, respectively, then

$$x = X + h$$
$$y = Y + d.$$

If the variables were measured without error, the regression slope would be B, whereas the regression using x and y produces an estimated coefficient b:

$$\hat{y} = a + bx.$$

The relationship between the true coefficient B and the estimated coefficient b is

$$b = B \left[1 + \frac{R_{Yd}\sigma_d}{\sigma_Y} \right] \left[\frac{1}{1 + \frac{\sigma^2_h + R_{Xh}\sigma_X\sigma_h}{\sigma^2_X + R_{Xh}\sigma_X\sigma_h}} \right],$$

where the σ's are the standard deviations of the respective random variables and the R's are the correlation coefficients. The first bias term in brackets arises from the error in measuring Y. If this error is always the same ($\sigma_d = 0$) and / or the error is not associated with the value of Y ($R_{Yd} = 0$), then the error in measuring Y contributes no bias to the estimate of the regression slope. The second bias term in brackets arises from the error in measuring X. There would have to be a very fortunate accident (R_{Xh} exactly equals $-\sigma_h/\sigma_X$) for this bias term to disappear; its usual effect will be to make the regression coefficient b an underestimate of the true value B. This means that the eager analyst, keen to discover and share useful new relationships among variables, will often be disappointed, finding associations to be weak when they should be strong.

The formula for b suggests three ways to cope with this problem of biased coefficients. One is to minimize the size of the fluctuations produced by measurement error (minimize σ_h). The second is to obtain data over a wide range of the variable X (maximize σ_X). The third is to obtain separate estimates of σ_h and R_{Xh} and use these in the formula for b to determine a new estimate of the true coefficient:

$$B = b \left[1 + \frac{\sigma^2_h + R_{Xh}\sigma_X\sigma_h}{\sigma^2_X + R_{Xh}\sigma_X\sigma_h} \right] \bigg/ \left[1 + R_{Yd}\frac{\sigma_d}{\sigma_Y} \right].$$

Other techniques, such as the method of instrumental variables, are also available to help cope with measurement error but cannot be discussed here (see Kmenta's book).

Because measurement problems are so common in planning applications, you should stay alert to the possibility that regression coefficients of small size might really be larger were it not for the measurement error. Thus negative results from regression analysis are not conclusive evidence against the existence of causal links between variables, just as positive results need not mean that there is a causal link.

Outliers

The final point to be made about mistrusting the numerical value of a regression coefficient concerns the influence of small numbers of cases having unusual values of the variables—*outliers*. The least-squares criterion for fitting gives special emphasis to points far from the regression line: a residual 3 times as big as another will contribute $3^2 = 9$ times as much to the sum of squared residuals and therefore have a far greater

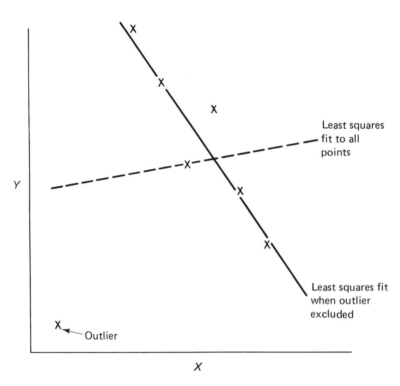

Figure 10.14
The influence of an outlier on a least squares fit

impact on the fitting of the line. If our regression were meant to summarize a relationship between two variables, we might not want one or a few odd cases to dominate the summary. For instance, consider a scattergram like that in figure 10.14. The least-squares fit to all the points works very poorly and misses the strong inverse relationship between X and Y. Excluding the single outlier from the analysis produces a fit that does much better at summarizing the relationship. Computing the correlation coefficient or the regression line blindly, without bothering to look at the scattergram, would be seriously misleading.

How do we identify outliers, and how do we deal with them? In the bivariate case it is sometimes possible to detect outliers visually, as in figure 10.14, although even in the bivariate case it is often difficult to spot any but the most extreme outliers. In the multivariate case it is usually hopeless to try to identify outliers by looking at scattergrams; even if we could hope to succeed, we would run the risk of arbitrarily labeling as an outlier any point that seemed to break with one of the many ghostly straight lines our imaginations can impose on a nondescript cloud of data points. (Hoaglin and Welsch provide a method for systematically identifying cases with high influence, but professional judgment should be as important as any statistical procedure.) If you decide to exclude certain cases from an analysis, be sure to document the exclusion and your rationale for it. To the extent that some cases become deviant because of errors in recording or keypunching, you can deal with outliers by double-checking the values of data for suspicious cases. Finally, one of the most active areas in current regression research is the development of robust regression methods that avoid the sensitivity of simple least-squares fits to outliers by employing alternative fitting procedures; Mosteller and Tukey's excellent text describes these methods after presenting a very useful discussion of the general problems of reading regression coefficients.

Summary

The prediction errors made in a bivariate regression analysis become the objects of further stages of analysis. Good estimates of these errors can serve as correction factors or adjustments that improve the original bivariate predictions. Multivariate regression can be thought of as a sequence of bivariate analyses in which the residuals from the previous analysis are estimated from another set of residuals representing the fresh information contained in a new predicting variable. These latter residuals are formed after using all the previously included predictor variables to estimate the newest predictor variable.

It is not necessary to restrict multivariate regression equations to additive combinations of metric predicting variables. A nominal or ordinal variable can be incorporated as a predictor by converting it to a set of dummy variables, which take on only the values 0 or 1.0 and number one fewer than the number of discrete categories in the nominal or ordinal variable. Multiplicative models can be used by exploiting the logarithmic transformation; such models imply qualitatively different relationships among the predicting variables than do additive models.

The predictive value of multivariate regression relative to the sample mean is summarized by the coefficient of determination R^2. It is possible for the analyst to choose predicting variables that have no association with the predicted variable in the population but nevertheless are accidently associated in the sample. In such cases the value of R^2 can be misleadingly large. The methods of testing the statistical significance of an association in contingency tables can be used, with suitable modification, to test the statistical significance of the observed value of R^2. The null hypothesis of independence is that the true value of $R^2 = 0$, with any observed nonzero value of R^2 attributable to sampling accident. The test statistic is

$$F_{obs} = \frac{(R^2/k)}{(1 - R^2)/(n - k - 1)},$$

where n = number of cases and k = number of predictor variables. This test statistic has the F distribution with $df = (k, n - k - 1)$, when the null hypothesis of independence is true. The descriptive level of significance is

$$\alpha = \text{Prob} [F \geq F_{obs} | R^2 = 0 \text{ in population}].$$

A statistically significant value of R^2 does not necessarily mean that the regression analysis predicts well enough for practical purposes, only that it outperforms the sample mean. The typical size of a prediction error is given by the *SEE*.

Use of the F test for statistical significance testing (and use of Student's t distribution for creating highest density regions) requires that certain assumptions hold. The regression residuals should have a Gaussian distribution with zero mean and constant variance (homoscedasticity) throughout the range of the predicted variable, and successive residuals should be uncorrelated. Plotting the residuals on probability paper provides a check of the reasonableness of the Gaussian assumption. Plotting the residuals against their associated predicted values and looking for trends in central tendency and dispersion provide a check of the assumptions of zero mean and homoscedasticity, respectively. Plotting the residuals against their successors (or determining the value of the Durbin-Watson statistic) provides a check against autocorrelation. These graphical techniques can be supplemented as necessary with more formal techniques.

Since many planning studies rely on observational rather than experimental data, and since most planning problems involve several variables, multivariate regression is one of the most frequently used methods for developing or testing causal theories. Unfortunately, the usual interpretation of a regression coefficient as "the change in the dependent variable caused by a unit change in the independent variable when all other independent variables are held constant" may not be a good response to the planner's desire for causal knowledge. First, the situation in practice may call for knowing what happens when all other variables react to the change in the one being manipulated, rather than knowing what happens when all other variables are held constant. Second, the association evident in a regression analysis may be predictive but noncausal. Third, the numerical values of the regression coefficients may be uncertain or biased if the data are flawed by reason of extreme multicollinearity, errors in measurement, or presence of outliers.

Multicollinearity, always present to some degree in any sample, inflates the standard errors of regression coefficients, resulting in uncertain estimates of their values. Errors in measurement of the regression variables, which often arise in planning studies when theoretical constructs are imperfectly operationalized, bias the estimates of regression coefficients, usually leading to underestimates of the effects. Truly deviant cases or cases made into outliers by errors in data processing also bias regression coefficients. These problems with the numerical values of regression coefficients can be minimized by gathering more data, especially for extreme values of the predicting variables, by using better measures of the variables, by taking care in data processing, and by using robust regression methods or identifying and setting aside deviant cases.

References and Readings

Andrews, D., and D. Pregibon. "Finding the Outliers that Matter." *Journal of the Royal Statistical Society* (B) 40 (1978): 85–93.

Armstrong, R., and M. Kung. "Least Absolute Value Estimates for a Simple Linear Regression Problem." *Applied Statistics* 27 (1978): 363–366.

Blodgett, J. "Obstacles to Accurate Statistics." *Publications of the American Statistical Association* 6 (1898): 1–19.

Bridges, W., and J. Oppenheim. "Racial Discrimination in Chicago's Storefront Banks." *Evaluation Quarterly* 1 (1977): 159–172.

Cochran, W. "Some Effects of Errors of Measurement on Linear Regression." *Proceedings of the Sixth Berkeley Symposium on Mathematical Statistics and Probability*, vol. 1. Ed. by L. M. LeCam, J. Neyman, and E. L. Scott. Berkeley: University of California Press, 1972.

Finkelstein, M. "Regression Models in Administrative Proceedings." *Harvard Law Review* 86 (1973): 1442–1475.

Hoaglin, D., and R. Welsch. "The Hat Matrix in Regression and ANOVA." *The American Statistician* 32 (1978): 17–22.

Kmenta, J. *Elements of Econometrics.* New York: Macmillan, 1971.

Mosteller and Tukey. "Regression for Fitting," chapter 12, "Woes of Regression Coefficients," chapter 13, and "A Class of Mechanisms for Fitting," chapter 14, pp. 259–297, 299–332, 333–379.

Tufte. "Introduction to Data Analysis," chapter 1, and "Multiple Regression," chapter 4, pp. 1–30, 135–163.

Problems

10.1

Mera studied the relationship between urban concentration and economic development in less developed countries (LDC's). He reasoned that if efficiency is increased by increasing urban concentration, then those LDC's with the greatest increases in concentration should also have the fastest growing economies, all else equal. Mera measured concentration in terms of "primacy," defined as the share of a nation's population living in the largest city (or in the largest two or three cities). He obtained data for the 47 LDC's with 1960 populations above 1 million on the change in primacy (ΔP) over a seven-year period and on the change in the per capita gross domestic product (GDP).

a. After setting aside Libya, whose GDP grew at a meteoric 21.4 percent rate after discovery of oil deposits, Mera regressed the rate of growth r on the change in primacy of the largest city ΔP_1:

$$\hat{r} = 1.763 + 0.411\Delta P_1$$
$$R^2 = 0.088.$$

Is this a statistically significant result?

b. Mera next repeated the analysis for only those countries with populations above 10 million. The result for the 19 cases was

$$\hat{r} = 1.305 + 0.859\Delta P_1$$
$$R^2 = 0.253.$$

Mera attributed the improved predictive power in part to the relative invulnerability of larger nations to accidental developments that cause the primacy to fluctuate over time. Interpret this explanation in terms of errors in measurement of variables.

c. What other explanation might there be for the increase in slope, and how might you test it?

d. Finally, Mera attempted to isolate the impacts of changes in concentration in the largest city (ΔP_1) from changes in the second and third largest cities ($\Delta P_{2,3}$). Among the 21 nations for which data were available, he found

$$\hat{r} = 1.428 + 0.454\Delta P_1 + 1.229\Delta P_{2,3}$$
$$R^2 = 0.237.$$

Does this equation predict significantly better than the sample mean?

e. Do Mera's statistically significant results prove that urban concentration leads to rapid growth?

10.2

In an exploratory study, Willemain examined the relationship between neighborhood variety and population mobility in a sample of 200 neighborhoods. Mobility was indexed by the percentage of the neighborhood population aged five years or over who had lived in the same house for five or more years (*NOTMOVER*) and by the percentage who had lived in a different county at some time during the previous five years (*CTYMOVER*). Variety was measured by the entropy of discrete distributions having three or four categories of the following variables: age of neighborhood population (*UAGE*), education level of working-age adults (*UEDUC*), income (*UINCOME*), value of housing (*UVALUE*), level of rent (*URENT*), rent-to-income ratio (*URENTINC*), number of dwelling units per building (*UUNIT*), and age of housing stock (*UHOMEAGE*). Correlations among these variables and their means and standard deviations are presented in the table that follows.

a. Estimate for the population age five years or older who moved within the previous five years, the percentage whose moves were within the same county.

b. What value of the correlation coefficient R corresponds to a 5 percent level of significance in a sample of 200 cases?

c. Would either *NOTMOVER* or *CTYMOVER* seem especially appropriate for a logit transformation?

d. Discuss the relative prospects for an analysis of *NOTMOVER* to return well-defined estimates of the regression coefficients for these two pairs of predictor variables: *UINCOME* and *URENTINC*, *UHOMEAGE* and *UVALUE*.

Correlation

Variable	NOTMOVER	CTYMOVER	UAGE	UEDUC	UINCOME	UVALUE	URENT	URENTINC	UUNIT	UHOMEAGE	Mean	Standard deviation
NOTMOVER	1.00	−0.82	0.56	−0.05	−0.27	0.16	0.02	−0.28	−0.10	0.17	57.1	11.2
CTYMOVER		1.00	−0.59	−0.20	0.21	−0.05	0.01	0.19	−0.08	0.01	16.9	11.1
UAGE			1.00	0.17	−0.06	0.26	0.04	−0.08	0.12	0.01	0.40	0.03
UEDUC				1.00	0.13	0.04	0.02	0.01	0.40	−0.20	0.47	0.04
UINCOME					1.00	0.00	0.14	0.29	0.11	−0.22	0.37	0.03
UVALUE						1.00	−0.03	−0.11	−0.06	0.20	0.29	0.08
URENT							1.00	0.22	−0.08	0.23	0.28	0.08
URENTINC								1.00	0.15	−0.22	0.37	0.06
UUNIT									1.00	−0.28	0.30	0.11
UHOMEAGE										1.00	0.36	0.11

10.3

Five of the variety measures with the highest correlations with *NOT-MOVER*, together with the median family income (*MDFAMINC*), were used to predict the value of *NOTMOVER* in each neighborhood. The results of this analysis are presented in the table.

a. Which regression coefficients are not clearly distinguished from zero, if any?

b. Analysis of the standardized residuals revealed the following distribution:

Range of
standardized residual: below -1 -1 to 0 0 to $+1$ above $+1$
Number of
residuals within range: 11   68   121   0.

What are the implications for statistical inference?

c. What kinds of neighborhoods appear to be especially stable?

Variable	Coefficient	Standard error of coefficient
UHOMEAGE	3.671666	6.24946
UAGE	182.2570	18.62941
UINCOME	-66.79589	21.21278
MDFAMINC	0.722498	0.31419
UVALUE	1.435877	7.86406
URENTINC	-30.17877	10.91394
(Constant)	11.72321	(Not computed)
$R^2 = 0.42427$		
$SEE = 8.59167$		

10.4

The mobility variable $CTYMOVER$ was predicted from median family income ($MDFAMINC$), median years of school completed by working age adults ($MDSCHOOL$), and the four variety measures having the highest bivariate correlations with $CTYMOVER$. The results of this analysis are presented in the table.

a. Compare the bivariate and multivariate analyses in terms of the predictive power of the mix of education levels in the neighborhood. (Hint: since the bivariate correlation coefficient is the coefficient in a regression of standardized variables, determine the value of the standardized multiple regression coefficient for $UEDUC$, with the help of the standard deviations listed in problem 10.2.)

b. How well can the value of $UEDUC$ be predicted from the rest of the predictor variables? Does multicollinearity explain the poor estimate of the coefficient of $UEDUC$?

c. What sort of neighborhoods appear to be especially stable according to the index $CTYMOVER$?

Variable	Coefficient	Standard error of coefficient
$UAGE$	−151.7781	18.72363
$UINCOME$	36.57854	21.32721
$MDFAMINC$	−0.67805	0.33502
$MDSCHOOL$	3.327203	0.85862
$UEDUC$	−15.38262	15.46293
$URENTINC$	13.55982	10.60231
(Constant)	32.39137	(Not computed)
$R^2 = 0.44141$		
$SEE = 8.39004$		

IV MAKING COMPARISONS

11 Experiments and Quasi-Experiments

A major planning activity is the assessment of change in policy or program. These comparative assessments are so fundamental to the reform of practice that it is essential to develop structured ways of thinking about them. There are a set of ideas that can organize your reaction to other people's comparisons and assist the planning of your own. For simplicity we will consider only the comparison of two alternatives: for example, housing quality in units with and without rent control, or before and after rent control; recidivism rates among juvenile offenders with and without pre-release training; quality of medical care provided in neighborhood health centers compared to hospital clinics; air quality levels before and after a new set of enforcement procedures; retail trade volume before and after conversion of a downtown area into a pedestrian mall; income level of high school dropouts with and without special job training. Many interesting comparisons involve two alternatives; the same basic considerations apply to comparisons of multiple alternatives as well.

There are several reasons for you to become tough-minded about such comparisons. First, if you must rely on comparisons made by others, you should be able to assess their validity. Second, if a program of yours is subjected to a critical but invalid evaluation, you will want to be able to defend yourself. Third, if you design a comparison yourself, you will want to do it as well as you can and to be aware of its limitations. To accomplish any of these ends you must learn something about the structure of comparisons and the litany of ills to which they are heir.

You should assume a posture of constructive skepticism. Just about any comparison will reveal some difference between the two programs being compared. What matters is whether the difference in result is actually attributable to the difference in programs, whether the difference can be counted on to reappear in other places at other times, and whether the difference is large enough to impact on practice. The first two issues are the province of experimental design and involve the internal and external validity of comparisons.

Internal Validity

The question of internal validity concerns whether the observed difference between the groups has some plausible explanation other than their difference in treatment. Sometimes the inability to attribute results arises from a simultaneity of changes in programs and environment, either of which might be responsible for a shift from the *status quo ante*. For instance, suppose a study of rent controls shows a serious decline in the quality of rent-controlled units after adoption of rent control, but this occurs in a city with a strong prior trend toward deterioration of the housing stock. One might argue with equal force either that rent controls result in decay of the housing stock or that rent controls have no impact on the quality of housing, attributing the deterioration to the continuation of the trend. The circumstances of such an evaluation preclude a firm assessment of the hypothesis "rent controls lead to housing decay" (if one had a suitably sensitive measure of housing quality, one might be able to investigate whether rent controls accelerate or decelerate the trend toward deterioration). The problem in this case is that ongoing processes rather than the new policy may be responsible for any changes observed over time. As a second example, note that either improved emergency medical services or the imposition of a 55 mile per hour speed limit might be responsible for a drop in the highway death rate; only a very detailed analysis would have a chance of isolating the impact of improvements in emergency medical services made at the same time that the speed limit is reduced. *Plausible rival hypotheses* are devised to contest the hypothesis that actions taken by the planner are responsible for the difference.

Sometimes the plausible rival hypothesis concerns differences between two groups receiving different treatments. In the case of "pet projects" one must be wary of stacking the odds in favor of an innovation by choosing only the best cases to receive it. For instance, a prison work-release program may appear to be successful, but is the success attributable to the positive impact of the program or to a prisoner selection process that allows only model prisoners to participate? In the case of "desperation projects" one must watch for the opposite kind of bias. Are death rates higher in university medical centers than in community hospitals because the quality of care drops when medical students and interns are involved or because the most severely ill patients are routed to the medical centers?

The work of Campbell and Campbell and Stanley provides a systematic list of nine types of rival hypotheses. In any particular comparison, one or more of these may provide a credible challenge to the view that a difference in programmatic outcome can be attributed to the experimental treatment. You should make it a practice to force any comparison to run the gauntlet of these rival hypotheses before concluding that the planned change deserves credit (or blame) for the difference in results.

1. *History*. This is the threat to contest the planner's hypothesis in the highway death rate example. Between before and after measurements some external factor besides the experimental treatment may arise to influence the variable of interest. In a physics or chemistry lab it is relatively easy to control external events, but who can say when New York City will suffer its next power blackout? When American Motors will close or open a plant? When news of municipal corruption will appear? Such events may have major influences on variables of interest in a comparison. Although we may not be aware of the significance of all events at the time they occur, we should be alert for some that could obviously confound interpretation of the data should they occur.

2. *Maturation*. Factors internal to the subjects of the experiment may be evolving in such a way as to confound interpretation of the results. This is the threat to internal validity in the rent control example, in which the quality of the housing stock was already decaying throughout the course of the comparison. As another example, note that a job-training program will appear more successful if it is implemented during a period of general business expansion.

3. *Selection*. This is the threat to internal validity in the prison work-release and medical center examples. In a with-and-without comparison where only those best (worse) suited for success receive the experimental treatment, the treatment may appear to be very good (bad) even if it actually has no effect at all. If a job-training program requires that its students apply for enrollment, pass screening tests for aptitude, and attend classes regularly, its graduates are very likely to be more successful even if the training program has added nothing to their skills, since the graduates comprise a select group with preexisting qualities commending them to employers.

4. *Selection-maturation interaction*. Comparison groups may be created in such a way that the natural rates of change in the two groups would be different even in the absence of the experimental treatment. For instance,

if you wanted to test the utility of a certain type of zoning for moderating growth, you should not group cities in such a way that all those with high growth potential (good transportation facilities, nearby recreation, large oil deposits close by) comprise one group.

5. *Experimental mortality*. Sometimes cases, be they individuals, families, neighborhoods, or cities, will drop out of an experiment before it is completed. If attrition proceeds unevenly in the two groups, it can be misleading to analyze only the cases that remain. For instance, imagine a study of neighborhood attitudes toward half-way houses for drug addicts. Some neighborhoods with half-way houses may experience well-publicized episodes of crime traced to those living in the half-way houses and thus manage to have those half-way houses closed. What happens to the final comparison of neighborhood attitudes toward half-way houses? Those neighborhoods with the strongest feelings against half-way houses have removed themselves from the study, so the remaining neighborhoods will represent a group biased in favor of half-way houses relative to the full set of neighborhoods initially in the study.

6. *Regression artifacts*. The word "regression" here refers not to the curve-fitting technique but to the tendency of extremely high or low values of some variable to be followed on next measurement by more moderate values (regression toward the mean). Suppose that commuter congestion is extremely heavy one week solely because of a random fluctuation in the number of travelers. Suppose further that the city traffic department changes the timing of the traffic lights to improve traffic flow. Even if the change in the lights has no effect, it is possible that the next week's traffic conditions will be better just because the peak congestion level has no place to go but down. Whenever we respond to an extreme value of some variable of interest (traffic flow, crime rate, and so on) *and* the extreme value was merely a random fluctuation un-correlated with other fluctuations, it is likely that the variable will take on a more moderate value at the next measurement even without our help. The issue is whether we can attribute an observed improvement to our intervention or simply call it a regression artifact.

7. *Instrumentation*. Another possible confounding factor is that the in-struments of measurement may differ. For instance, interviewers may be more comfortable, perceptive, and knowledgeable at a second interview

than a first, or two different raters of housing quality, both trained on the same set of slides, may still disagree as to whether a given house is "dilapidated". We must strive for clarity of definition and consistency of measurement. There will almost always be random variations in measurement, but these may not concern us too greatly if they are well mixed between the two groups. On the other hand, systematic differences in measurement may severely bias the comparison as, for instance, when one of the two groups of houses is judged exclusively by the rater more inclined to label a house "dilapidated."

8. *Testing.* Whereas instrumentation refers to those doing the measuring, testing refers to changes induced in those being measured by the very process of measurement. For instance, a questionnaire respondent may be better oriented to the "after" than to the "before" questionnaire, or an interviewee may be more forthcoming during a second interview than a first. The term derives from educational evaluations, in which an improvement in test scores after a unit of instruction might be as much due to learning better how to take the test as to learning the subject matter.

9. *Instability.* We know that variables take values that fluctuate over time and that sample results differ to some extent from sample to sample. It may be that the difference between before-and-after or with-and-without is attributable solely to these chance fluctuations. If the experiment were repeated, the difference might disappear or even change sign. Larger numbers of cases and the mechanisms of statistical inference help deal with this threat.

These nine threats to internal validity comprise a useful checklist for critical scrutiny of any comparison. Some are present to at least a small degree in any comparison, but it is not their general existence but their salience in any particular situation that matters most to the planner. As we shall see, comparisons can be structured in ways that effectively render any of these threats implausible, although this protection from rival hypotheses comes at the price of more elaborate and expensive comparisons. It may help to summarize the nine threats into four general classes:

1. *Other influences.* Some event or force other than the experimental treatment may produce the results observed (history, maturation).

2. *Biases*. The choice of individuals to constitute the two groups may lead to pseudo-differences caused by unfair comparisons (selection, selection-maturation interaction, experimental mortality).

3. *Measurement*. Factors in the measurement process may lead to pseudo-differences (instrumentation, testing).

4. *Randomness*. The observed difference may be a statistical artifact (instability, regression artifact).

Later in this chapter we will introduce a mathematical treatment of these threats to internal validity that should help your understanding of how they work and how they can be countered by properly designing the comparison.

External Validity

While questions of internal validity focus on whether one can draw a relatively unambiguous conclusion about the effect of an experimental treatment, questions of external validity focus on the generalizability of the experimental results to the rough-and-tumble world of practice. Threats to external validity can be grouped into two classes:

1. *Artificiality*. The reactions of subjects in experiments may not be representative of the reactions of similar subjects in nonexperimental settings. First, the very fact that subjects (individuals, hospitals, cities, and whatever else) agree to participate in a study—even if they do not receive the experimental treatment, serving only as a reference or *control* group— may mark them as unique (progressive? compliant? desperate? inquisitive?). Second, even if those studied are typical of the larger population of interest, their behavior under observation may differ a good deal from their normal behavior (the *Hawthorne effect*).

2. *Replicability*. There may be difficulty reproducing good results in other settings. First, the care, talent, and money invested in an experimental program may not be available to support widespread implementation of the program. Second, even a simple treatment has many components; only one or a few may be responsible for positive impacts, but those may be just the details missing in another setting (for example, the charismatic community leader)—one doesn't know without more detailed investiga-

tion which parts of the treatment bundle are critical, but the bundle will change from site to site.

Unfortunately, it is possible to get in a position where an attempt to improve the internal validity of a comparison may jeopardize its external validity, and vice versa. For instance, it may be useful from the perspective of internal validity to obtain large quantities of "before" data to develop a solid baseline for later reference in a study of the arrest patterns of policemen soon to be enrolled in community-relations classes. However, the sudden appearance of earnest, bespectacled research assistants swarming around files and muttering about use of force in certain areas of town cannot help but forewarn the subjects of the study, who may well react in anticipation of new policy directives. To the extent that new programs or policies are evaluated under conditions that resemble a laboratory, the care in execution and the control of confounding factors achieved in the evaluation may move the results rather far from the world of practice. However, if the evaluation is internally valid and the results positive, there can then be little doubt of the value of further studies to explore external validity.

Only rarely will you be able to avoid the tension between internal and external validity when you design comparisons. Still, you can count on some help from the fact that not every threat to internal or external validity is credible in every situation; only those threats plausible in your particular case must be dealt with, so running a theoretical risk may be running no risk at all. Your antagonists can be counted on to point out the relevant theoretical risks while you are developing your sense of the plausibility of rival hypotheses and afterwards. A much greater concern may well be to convince your colleagues that the programs that seem so obviously desirable a priori really ought to be carefully evaluated. It is less difficult to stir up a debate about the methodological soundness of a comparison than to prevail on those responsible for policies and services to adopt an experimental mentality in the first place.

Structures for Comparisons

Now that we have reviewed the various threats to internal and external validity, we can study several different experimental designs and assess them in terms of their ability to fend off these threats. In the best tradition of sport, we will use a system of X's and O's to notate our strategies. An O represents an observation (measurement) of a variable of interest and an X represents an application of the experimental treatment. Time flows from left to right. Thus the design

$$O \quad X \quad O$$

represents a before-and-after comparison having a before observation (often called a "pretest") followed by the application of the new policy or program (the experimental X), then an after (or "posttest") observation. Likewise the design

$$X \quad O$$
$$O$$

represents a with-and-without comparison involving both a study (or experimental) group, which receives the experimental treatment and is then observed, and a control group, which receives something other than the experimental treatment (perhaps nothing, perhaps the status quo, or perhaps even a second innovation) and is then observed at the same time as the study group. A critical feature of any comparison involving both a study and a control group is how individual cases are assigned to each group. If they are assigned randomly, we will affix an R to the diagram

$$R \quad \begin{array}{cc} X & O \\ & O \end{array}$$

There are many conceivable designs for comparisons. Some of them can be arranged in a kind of evolutional hierarchy as shown in figure 11.1. Following Campbell and Stanley, we group the designs into three classes: pre-experiments, quasi-experiments and true experiments. A true experiment requires random assignment of subjects to the study and

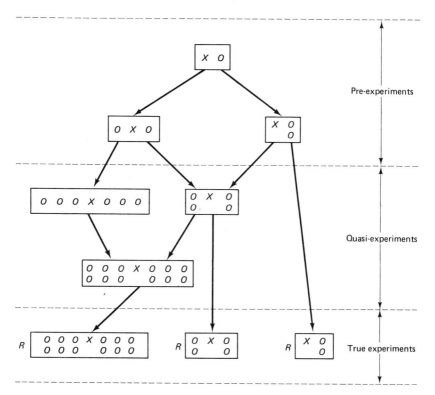

Figure 11.1
Evolutionary hierarchy of experimental designs

control groups. Simpler designs become more complex, more costly, and more internally valid as more information and more experimental control are invested.

The Demonstration

We call the pre-experimental design

$$X \quad O$$

a *demonstration*, wherein some policy or program is tried, then observations are made. Strict constructionists scoff at this primitive structure, since without explicit reference data no formal comparison of alternatives can be made. Nevertheless, such studies are very common, and their prevalence does not derive solely from the presumed ignorance of practitioners. While a demonstration cannot be used to compare two alternative programs, it can be used to show that it is indeed possible to implement the innovation, to test for catastrophes in the execution of the treatment (Was the public outraged? Did the staff threaten to quit? Were there lawsuits? Did anyone abscond with funds?), and to see if pre-established performance norms can be met ("We'll consider your innovation further if you can prove that your new system can handle the present workload.") The demonstration therefore has much to commend it as a preliminary step toward adoption of a new program. Unfortunately, it can also serve as a smokescreen for those afraid to really test competing approaches or for those so committed to a course of action that the demonstration substitutes for a research program. If we care at all about client outcome rather than the mere appearance of action of their behalf, we cannot quit with a demonstration, although occasionally we may well begin with one.

The Before-and-After Design

We have called the pre-experimental design

$$O \quad X \quad O$$

a *before-and-after* design. This design is well suited to the physical sciences, where laboratory isolation and control can be well maintained. Its major vulnerability is to confounding by other factors—either history or maturation or both could be the major cause of any differences observed between the before and after measurements. To some extent one can limit the possible confounding influences of history by keeping the interval between observations brief, while still allowing enough time to apply the treatment and let it take effect. The fact that subjects and researchers can learn from the before observation opens the design to the threats involving measurement—instrumentation and testing.

The With-and-Without Design

We have called the pre-experimental design

a *with-and-without* design, wherein the study group receives the experimental treatment, the control group does not, then simultaneous observations are made of both groups on the variables of interest. Note that this design offers some protection against history to the extent that whatever external forces influence the study group also influence the control group. For instance, a change in mayor, governor, or president might greatly influence a before-and-after study but affect the study and control groups equally in a with-and-without design, provided both groups are in the same political jurisdiction. Since only the *differences* between study and control are of interest, the change in administration should not confound the interpretation of results. To the extent that the subjects in the groups are similar, the with-and-without design also protects against maturation. However, it is quite vulnerable to selection biases, since it is very difficult to insure the comparability of the study and control groups in the absence of random assignment of cases to groups. Since the with-and-without study measures each group only once, it offers some protection against the instrumentation and testing threats to internal validity. In terms of external validity, it has the advantage that neither study nor control group is pretested, so some experimental artificiality can be avoided.

The Nonequivalent Control Group Design (Jack Sprat)

This is a good quasi-experimental design with a somewhat ungainly name. It is a combination of the before-and-after and with-and-without designs and therefore combines the best features of each. Since those two designs have complementary strengths and weaknesses, I find it helpful to refer to their combination as the *Jack Sprat* design after the children's verse:

Jack Sprat could eat no fat.
His wife could eat no lean;
And so betwixt the two of them
they licked the platter clean.

Because the strengths and weaknesses of the before-and-after and with-and-without designs are quite complementary, merging them provides a good deal of protection against threats to internal validity. The with-and-without component helps protect against history, maturation, and measurement problems, while the before-and-after component helps protect against selection biases.

A mathematical model of the Jack Sprat design best explains the weaknesses of the simple, pre-experimental designs contained within it. Suppose we are interested in a metric variable, perhaps the annual income of an individual, the full-value tax rate of a community, the death rate of a hospital, or the support for the mayor in a neighborhood. Let us also assume that we are studying the benefits of some particular intervention, perhaps a job-training program for the individuals, a management innovation for the community, a new quality review audit for the hospital, or a public relations initiative for the neighborhood. We label the observations as follows:

$$\begin{array}{ccc} O_1 & X & O_2 \\ O_3 & & O_4 \end{array}$$

For simplicity, we will work only with average values within each of the two groups, such that

O_1 = pretest mean for study group on variable of interest,
O_2 = posttest mean for study group on variable of interest,
O_3 = pretest mean for control group on variable of interest,
O_4 = posttest mean for control group on variable of interest,
X = mean change in variable of interest caused by experimental treatment,

H = mean change caused by history in study group,
h = mean change caused by history in control group,
M = mean change caused by maturation in study group,
m = mean change caused by maturation in control group,
I = mean change caused by instrumentation in study group,
i = mean change caused by instrumentation in control group,
T = mean change caused by testing in study group,
t = mean change caused by testing in control group.

A simple additive model will illustrate the confounding effects of history, maturation, selection, selection-maturation interaction, instrumentation, and testing. Since we are working with group means, we will assume that the averaging process eliminates instability as a serious threat to internal validity. For simplicity we will ignore the threats posed by regression artifacts and experimental mortality. Using the above definitions, we can relate the pretest and posttest means in each group as follows:

$$O_2 = O_1 + H + M + I + T + X$$
$$O_4 = O_3 + h + m + i + t.$$

Thus the mean value of the variable of interest among the study group after the intervention (O_2) is the value before the intervention (O_1) plus any change attributable to history (H), plus any change attributable to maturation (M), plus any change due to instrumentation or measurement error (I), plus any change attributable to testing (T), together with any change attributable to the experimental intervention (X). A similar relationship holds for the control group, except that there is no experimental impact X.

Consider the before-and-after component of the Jack Sprat design. If we were using only the study group data, we would estimate the experimental impact by computing the difference between the averages before and after the intervention, which we will call D_{BA}:

$$D_{BA} = O_2 - O_1$$
$$= [H + M + I + T] + X.$$

Because the bracketed term is made up of factors that clutter the interpretation of the comparison, I refer to it as the "junk term." If it were not under foot, the experimental effect could be clearly determined. Ideally, the term in brackets would be zero, so that the mean before-and-after difference would simply equal X, the average value of the experimental effect. However, what we actually measure is D_{BA}, and we cannot know how this difference is constituted. Suppose $D_{BA} = 11$; is the junk term $+14$ and the experimental impact -3, or vice versa, or some other combination totaling 11? The effects of history, maturation, and measurement error are inextricably linked with the experimental impact if we ignore the control group data

Now consider the with-and-without portion of the Jack Sprat design. If we were only using the posttest data from each group, we would estimate the impact of the experimental treatment by computing the difference between the mean posttest values of the variable of interest in both groups. Call this difference D_{WW}:

$$D_{WW} = O_2 - O_4$$
$$= [(O_1 - O_3) + (H - h) + (M - m) + (I - i) + (T - t)] + X.$$

Again the ideal situation would be for the junk term in brackets to equal zero. Since it is unlikely that all confounding factors will disappear, however, the real issue is whether the junk term is much smaller than X, the mean experimental effect.

There are two differences between the junk terms of the two constituent designs that are merged to form the Jack Sprat design. First, to its credit, the junk term of the with-and-without design is composed of subtracted pairs, so even if the individual factors (such as H, h, M, m) are not zero, they may effectively cancel if the two groups are sufficiently alike (perhaps $H - h \approx 0$, $M - m \approx 0$, and so on). Second, to its discredit, the junk term for the with-and-without design has an extra component, $O_1 - O_3$, which represents the selection bias between the study and control groups. Since we have no way of knowing whether the advantage of possible cancellations is offset by the disadvantage of possible selection bias, we cannot categorically state a preference for either the with-and-without design or the before-and-after.

Now, consider what happens if we make full use of the data available in the Jack Sprat design. The analysis of the Jack Sprat requires computing the *difference of differences*, which we will call D_{JS}:

$$D_{JS} = (O_2 - O_1) - (O_4 - O_3)$$
$$= [(H - h) + (M - m) + (I - i) + (T - t)] + X.$$

Note that the junk term for the Jack Sprat design looks better than either of the others: better than the before-and-after junk term because it provides for possible cancellation (although we have no guarantee that the cancellation will actually occur) and better than the with-and-without junk term because it contains no extra term reflecting selection bias.

To help summarize the comparison we juxtapose the estimates of experimental effect computed for the three designs:

$$D_{BA} = X + [H + M + I + T]$$
$$D_{WW} = X + [(H - h) + (M - m) + (I - i) + (T - t) + (O_1 - O_3)]$$
$$D_{JS} = X + [(H - h) + (M - m) + (I - i) + (T - t)].$$

If the two groups are similar in the sense that history, maturation, and measurement problems afflict them about equally, the junk term in the Jack Sprat design will be small, and D_{JS} will give a good estimate of the experimental impact X.

Randomization and Matching

The with-and-without design can be converted to a true experiment by randomly assigning subjects to the two groups:

$$R \begin{array}{|cc|} \hline X & O \\ & O \\ \hline \end{array}.$$

Likewise the Jack Sprat design can be converted to a true experiment by randomization:

$$R \begin{array}{|ccc|} \hline O & X & O \\ O & & O \\ \hline \end{array}.$$

We will refer to these as the "randomized with-and-without" and "randomized Jack Sprat" designs, respectively.

What good does randomization do? Consider the junk terms for both designs. Randomization simply makes it more likely that the potential cancellations occur so that the junk terms will in fact be small. This is especially true when there are large numbers of cases in each of the two groups. Randomization accomplishes this rough, statistical equivalence by breaking up any systematic patterns in the assignment of cases to the two groups.

There is continual debate about the ethics of randomization. Some say that if professionals have a strong feeling that a given intervention will be beneficial, they have no right to withhold that treatment from anyone. Others counter that the professional opinions must be suspect unless they arise from randomized comparisons. Some proponents of randomization exhort practitioners to try randomized comparisons, while other researchers devote years to trying to find approximate ways (by regression analysis, for example) to equate two groups in the absence of randomization. Another group tries to reconcile these differences by exploiting situations where randomization occurs naturally or can be introduced as part of a necessary process of rationing access to a limited treatment.

There can be no doubt that randomization of large numbers of subjects does improve the internal validity of a comparison. There can also be no doubt that the planner will not always be able to control access to the treatment to the extent required for randomization. Even where you can, your own ethical judgment will have to be made about randomization. If you do not randomize, you should give as little ground as possible, using good quasi-experimental designs like the Jack Sprat.

It is tempting to employ the technique of *matching* as a substitute for randomization. One identifies cases already in the study or control groups that are alike in one or several attributes believed to be important and compares them. For instance, in comparing two types of programs to stimulate economic development, one might single out for comparison only those cities in the 10 to 20,000 population range with median annual income above $8,000, reasoning that this will help guarantee a fair comparison.

However, any matching activity will inevitably undermatch, never really controlling for all relevant factors. The subsample of cities matched by population size and income will still differ in many important dimensions: perhaps their recent growth rates, tax rates, public service infrastructure, and annual rainfall will be stronger determinants of economic development than population and median income. If you concede this and try to match on these attributes too, you may end up with no two cities alike on all counts. More important, even if you could identify many apparently well-matched cities, there will be important unmeasurable (or hard to measure) attributes that may be unevenly distributed in the two groups. For instance, it may be that one of the development strategies requires a strong mix of both community consciousness and pro-business attitudes; cities opting for that strategy may fare differently just because of those very attitudes, which might not be found among the other group of cities. Thus we can never hope to match the two groups systematically on all attributes that matter. This does not mean that matching is useless, however. If randomization is not possible, matching does reduce the chance of misattribution to some (unknown) extent.

Furthermore, while it is not a good idea to substitute matching for randomization when the latter is possible, it is a good idea to combine the two. It is possible when randomizing relatively small numbers of subjects to produce outrageous imbalances between study and control groups: for instance, assigning nearly all the men to one group and nearly all the women to the other in a study of attitudes toward community day care, or assigning nearly all the young people to one group and nearly all the old to the other in a study of a dial-a-ride bus system. In such cases, we know on substantive grounds that the attribute so poorly mixed in the two groups plays a major role in influencing the variable under study, and we should not worship randomization to the extent of tolerating major imbalances between groups just because they arose by accident. Rather, a good procedure to follow is to identify pairs of cases well matched on key attributes and randomly assign one of each pair to the study group.

Monte Carlo Simulations

In earlier chapters we made use of synthetic data to develop a feel for the kinds of random fluctuations that arise in sampling problems. We can use the same basic approach to explore the behavior of alternative experimental designs. We will specify ahead of time the true magnitude of an experimental effect, then see how well the designs can recover the truth amidst such threats to internal validity as selection bias and instrumentation. Since this method of studying the performance of experimental design involves repeated random trials, the methodology is known as *Monte Carlo simulation.* It is commonly used whenever analytical approaches would be inconvenient or even impossible.

We will suppose we are evaluating the effectiveness of a job-training program. We will synthesize three items of information for each individual: initial earnings, a junk term that includes the effect of history (new plant openings), maturation (seniority, increases in the minimum wage), and measurement (willingness and ability to reveal accurate income information) and an experimental impact (which will by definition be 0 for someone assigned to the control group). An individual's final income level will be given by the sum:

Final income = Initial income + Junk term + Experimental impact,

with each term rounded to the nearest dollar. For the purposes of illustration, we will assume that each of the three items of information will be a random number drawn from a Gaussian distribution. Initial income will have mean $150 and standard deviation $50; the junk term will have mean $10 and standard deviation $10; the experimental impact will have mean $25 and standard deviation $5. Thus all income should appear to increase by about an average of $10 without the job-training program and by about $35 with it. The true experimental impact should appear to be about $25 on the average. Data for 25 simulated workers are listed in table 11.1. The table also indicates the group to which each worker is assigned.

The first simulated assignment rule is biased, in that workers having lower incomes are more likely to receive the experimental treatment. This biased assignment is accomplished using two standardized Gaussian numbers as follows. The first random number establishes a dividing line. If the second random number is greater than the first, the worker is assigned to the study group; otherwise, he is assigned to the control group. The worker's initial income level is determined by multiplying the first random number by 50, then adding 150. In this way, when the first random number is small and therefore the worker's initial income

Table 11.1
Data for simulation of experimental designs with biased assignment of workers to groups

Worker number	Group	Initial income	+	Junk term	+	Experimental impact	=	Final income
1	S	122		12		24		158
2	S	123		−2		21		142
3	C	187		16		—		203
4	C	278		−2		—		276
5	S	86		7		27		120
6	C	213		−1		—		212
7	S	112		15		35		162
8	S	93		9		23		125
9	S	62		0		33		95
10	S	147		6		35		188
11	S	89		8		23		120
12	C	143		5		—		148
13	C	149		8		—		157
14	S	223		5		27		255
15	S	118		−10		27		135
16	S	130		5		35		170
17	C	152		28		—		180
18	C	221		7		—		228
19	S	123		1		17		141
20	C	212		21		—		233
21	C	133		−4		—		129
22	S	123		−4		25		144
23	C	160		−14		—		146
24	C	219		2		—		221
25	S	159		8		22		189

Table 11.2
Analysis of data in table 11.1

$$O_1 = \$122 \quad X \quad O_2 = \$153$$
$$O_3 = \$188 \quad \quad O_4 = \$194$$

$$D_{BA} = O_2 - O_1 = \$31$$
$$D_{WW} = O_2 - O_4 = -\$41$$
$$D_{JS} = (O_2 - O_1) - (O_4 - O_3) = \$25$$
True value of $X = \$25$

is low, the worker has a greater chance of being assigned to the study group. (To generate a Gaussian random number with mean μ and standard deviation σ, multiply a standardized Gaussian random number by σ and then add μ.)

The results of this first simulation are displayed in table 11.2, which shows the mean initial and final incomes in each group and the measure of experimental impact computed three ways. If the data were from a before-and-after comparison (ignoring the 11 control group workers), the estimate of experimental impact would be $D_{BA} = \$31$; this is an overestimate caused by the junk term, which was expected to add an average of \$10 to each worker's income over the period of the experiment. On the other hand, if the data were from a with-and-without comparison (ignoring all 25 initial income levels), the selection bias against the study group leads to a serious underestimate of experimental impact: $D_{WW} = -\$41$. Finally, if all the data are used in a Jack Sprat design, the estimate of experimental impact happens in this case to be exactly equal to the theoretical mean effect: $D_{JS} = \$25$. While events need not always work out so nicely, these results indicate the value of investing in the more costly Jack Sprat design as a way of protecting against both maturation and selection bias.

A second simulation explores the value of randomly assigning workers to the two groups rather than generally assigning the poorest to the training program. The data for this second simulation are shown in table 11.3. Initial incomes, junk terms, and—where possible—experimental impact are unchanged from table 11.1. Results for the second simulation are shown in table 11.4.

Table 11.3
Data for simulation of experimental designs with random assignment of workers to groups

Worker number	Group	Initial income	+	Junk term	+	Experimental impact	=	Final income
1	S	122		12		24		158
2	S	123		−2		21		142
3	S	187		16		23		226
4	C	278		−2		—		276
5	S	86		7		27		120
6	C	213		−1		—		212
7	S	112		15		35		162
8	C	93		9		—		102
9	S	62		0		33		95
10	S	147		6		35		188
11	S	89		8		23		120
12	C	143		5		—		148
13	S	149		8		33		190
14	C	223		5		—		228
15	C	118		−10		—		108
16	S	130		5		35		170
17	C	152		28		—		180
18	C	221		7		—		228
19	C	123		1		—		124
20	C	212		21		—		233
21	S	133		−4		25		154
22	C	123		−4		—		119
23	S	160		−14		23		169
24	S	219		2		27		248
25	C	159		8		—		167

Table 11.4
Analysis of data in table 11.3

$O_1 = \$132$	X	$O_2 = \$165$
$O_3 = \$172$		$O_4 = \$177$

$$D_{BA} = O_2 - O_1 = \$33$$
$$D_{WW} = O_2 - O_4 = -\$12$$
$$D_{JS} = (O_2 - O_1) - (O_4 - O_3) = \$28$$
True value of $X = \$25$

We would expect no major differences to arise in the case of the before-and-after design, and in fact the estimate $D_{BA} = \$31$ is close to the previous figure of $33. On the other hand, we would expect randomization to improve greatly the with-and-without comparison by minimizing selection bias. Table 11.4 shows that much, but by no means all of the difference in initial group incomes is eliminated by the particular random assignment of cases shown in table 11.3. The initial difference of $66 has been reduced to $40, which is still a sizable bias and leads to the still very unsatisfactory estimate $D_{WW} = -\$12$. In this instance randomization has helped a little but not enough to avoid an improper conclusion about the benefits of the job-training program. *On average*, randomization will eliminate differences in initial income, but in any one particular case there can be no absolute guarantee that randomization will produce initially equivalent groups (although larger numbers of cases lead to greater assurance that randomization can help).

In the case of the Jack Sprat design, randomization will help only to the extent that the junk terms of the two groups differ, since selection bias is already controlled for. In our simulations we hypothesized the same distribution for the junk term in both groups, so randomization should make little difference in these simulations; the estimate $D_{JS} = \$28$ is close to the previous estimate of $25. Once again, the estimate provided by the Jack Sprat design is closest to the truth.

Given the degree of variability present in the hypothesized distributions of initial income, junk term, and experimental impact, it would be unwise to try to draw very firm conclusions from just two simulations. Generally, Monte Carlo simulations are repeated hundreds or even thousands of times before their results are summarized. You should not infer from these two simulations either that the before-and-after design is always preferable to the with-and-without or that randomization is generally ineffective in equating two groups. You should note that randomization is only effective on average, and that a quasi-experimental design like the Jack Sprat can be notably more effective at identifying experimental impacts than can a pre-experimental design. You should also have a more visceral sense of how uncertainty clouds our estimate of the good we might do when we reform or innovate.

The Time-Series Design

The quasi-experimental design

$$O \quad O \quad O \quad X \quad O \quad O \quad O$$

is called a *time-series design*. (The diagram shows three before and three after measurements solely for graphic convenience; in general, the more observations we can get the better.) This extension of the before-and-after design has obvious advantages. The longer look provided by the repeated observations before the experimental intervention helps in detecting maturation effects (trends). Repeated pretests also help shake-down the measurement process and acclimate the subjects of study to the process, thereby reducing the chances that instrumentation or testing effects will contaminate the data. The repeated measurements after the experimental intervention permit detection of two kinds of changes difficult or impossible to document with a before-and-after design: changes in trend (as distinct from changes in the average level of the variable of interest) and delayed impacts. The repeated observations also provide larger sample sizes for statistical inference and thereby lead to better estimates of effects. As in the before-and-after design repeated observations on the same cases mean that selection bias is not a serious threat to internal validity.

However, the time-series design is by no means flawless. It is perhaps more vulnerable than the before-and-after to certain problems of external validity. First, the fact that the subjects of the analysis are observed so often may make their responses relatively artificial (although it is sometimes possible that the observation be unobtrusive and thus lead to no significant changes in behavior). Second, cases that participate throughout the duration of the study may be unrepresentative of the larger population for just that reason. (They may be more docile, more interested, more tenacious, or less mobile then those not studied.) Once again we realize the tension that can arise between internal and external validity and are reminded that no design is perfect. Confounding by history remains a serious possible threat to internal validity in the time-series design.

Still, the net of these strengths and weaknesses is that the time-series design ranks with the Jack Sprat as a good, quasi-experimental design. Whereas the time series is vulnerable to history, Jack Sprat is vulnerable to selection-maturation interaction. Whereas the greater number of observations raises the possibility of artificial behavior in the time-series design, it also lessens the vulnerability to artifacts of instrumentation and testing. Both the time-series design and the Jack Sprat are superior to their pre-experimental predecessors, the before-and-after and with-and-without designs. Both could in turn be combined to form a more elaborate and costly design with hybrid vigor derived from the strengths of each; such a *multiple time series design* is illustrated in figure 11.1.

Statistical Inference in the Analysis of Comparisons

Of all the potential problems that may arise in the comparison of two programs, only the threat of instability is addressed by statistical methods per se. The structure of comparison is generally more important as a guarantee of valid and generalizable conclusions, and if the experimental innovation is dramatically successful, even a crude design and a rudimentary analysis will not obscure the fact of success. In short, statistical analysis is not the most important component of comparative program evaluation nor always even necessary.

Analysis does have a role, however. In most cases, successful new programs or policies will lead to marginal improvements rather than dramatic breakthroughs. These improvements may be large enough to be worth having but still small enough to require careful statistical analysis to document their existence and estimate their size. Furthermore, even large improvements may not be entirely obvious if there are only a few data points and a good deal of "noise" in the data. For these reasons the methods of statistical inference are useful and sometimes crucial adjuncts to the methods of experimental design.

The proper choice of statistical procedures will depend on the structure of the comparison and the level of measurement of the data. If the comparison involves matched pairs of observations, and the data are metric, then the familiar Bayesian analysis using Student's t test is recommended, provided the observed differences between paired values are independent and well described by the Gaussian distribution (refer to chapter 7). In the case of the before-and-after design, the differences are formed from the pretest and posttest scores of each individual in the study (in the notation introduced earlier in this chapter, the differences are computed as $O_2 - O_1$). In the case of the with-and-without design, the differences are formed from the observations made on each member of the matched pair following application of the experimental treatment to the study group (one value of $O_2 - O_4$ is computed for each matched pair). In the case of the Jack Sprat design, the differences are computed as the difference in changes from pretest to posttest within each pair (that is, $(O_2 - O_1) - (O_4 - O_3)$ is computed for each matched pair). If the differences are not well described as Gaussian, then the Wilcoxon signed rank test is a recommended alternative (see Mosteller and Rourke, chapter 5).

If the comparison involves two sets of observations but not matched pairs (or if it does but we ignore the matching), there are many alternative statistical procedures that might be applied. One of these alternatives— the *Mann-Whitney test*—is presented in the next section. It has advantages for planning applications in that its use requires neither large samples nor Gaussian or even metric data. All that is necessary is that the two sets of observations be independent and rank-ordered, so small sets of ordinal data can be analyzed with the Mann-Whitney test. Since planners often work with small data sets and subjective data that are ordinal at best, the Mann-Whitney test is well suited to the demands of practice. It also performs very well in less harsh circumstances.

The Mann-Whitney Test

The purpose of the Mann-Whitney test is to examine the plausibility of the hypothesis that an observed difference between two sets of comparative observations can be attributed to instability rather than experimental impact. Thus the test begins with the null hypothesis that the two sets of observations represent samples drawn from the same population distribution.

Which two sets of observations are to be compared depends on the choice of experimental design. For a before-and-after comparison, the set of before observations is compared to the set of after observations. For a with-and-without comparison, the study group scores are compared to the control group scores. With the Jack Sprat, the changes from pretest to posttest in the study group are compared to the corresponding changes in the control group.

If the data arise from a time-series design, the presence of trends in the variable of interest requires a somewhat different treatment. Suppose there is a steady increase in the variable of interest and the experimental intervention has no impact at all. This situation is depicted in figure 11.2. Note that simply dividing the observations into sets of before-and-after measurements produces two very distinctly resolved sets of points. It is very unlikely that these two sets of points would arise as samples from the same distribution, as any significance test would confirm. However, the reason that the two distributions are different is not the experimental impact (which we know in this case to be zero) but merely the uninterrupted trend. Thus a comparison of the original observations is not the equivalent of a test for experimental effect. On the other hand, working not with the original observations but rather with differences between adjacent observations gives a much more faithful depiction of the experimental impact. As shown in figure 11.2 the differences in the before period are indistinguishable from those in the after period. A test of the null hypothesis of no difference in the distribution of *changes* would properly conclude that there was no detectable experimental impact. Therefore the changes in the variable of interest should be compared, not the original levels of the variable. If the trend is exponential (like an inflation or population growth curve) rather than linear, you should remove the effects of trend by analyzing the ratios of adjacent observations instead of their differences.

To further illustrate the need to analyze differences (or ratios) rather than the original observations, consider the time series depicted in figure 11.3. In this case there is a dramatic experimental impact, with the increasing trend clearly reversed. However, an analysis based only on the

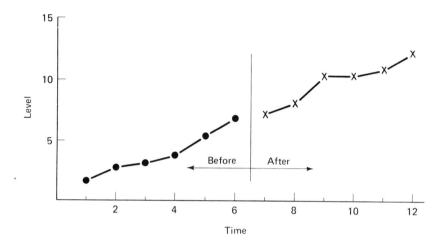

Distribution of levels

Distribution of changes in level

Figure 11.2
A case in which the distribution of changes in level is well mixed although the distribution
of levels is well resolved

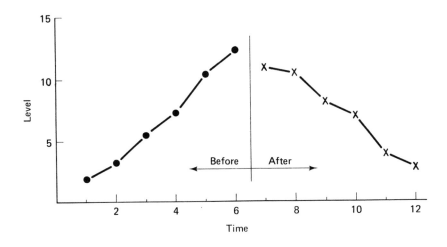

Figure 11.3
A case in which the distribution of changes in level is well resolved although the distribution of levels is well mixed

original observations would lead to a finding of no impact, since the before-and-after observations are so intermixed. On the other hand, the distribution of differences clearly shows two distinct groups and would lead to rejection of the null hypothesis of identical distributions and no experimental impact.

Whether we are working with two sets of original observations or two sets of differences, the Mann-Whitney test (or any other significance test) looks for evidence of experimental impact in terms of the resolution of the data into two distinct groups. Consider, for instance, the with-and-without comparison

$$X \quad O_2$$
$$\quad O_4$$

In terms of the additive model presented earlier

$$O_2 = [O_1 + H + M + I + T] + X = J + X$$
$$O_4 = [O_3 + h + m + i + t \,] = j,$$

where J and j represent the junk terms in the study and control groups, respectively. Now if the comparison is fair in the sense that J and j are random variables with the same distribution, and if there is no experimental impact, then the values of O_2 and O_4 should have the same distribution. Under these circumstances any observed separation of the values for the cases in the two groups must be attributable to random fluctuations. This is the perspective of the null hypothesis in the Mann-Whitney test: there should be much mixing of the values from the two groups.

Now suppose the null hypothesis becomes untenable because the two distributions resolve themselves more strongly than would be expected due to sampling accident. It is important to appreciate that in this case the conclusion from the significance test is ambiguous: either there was an experimental impact, or the comparison was unfair, or both. The significance test by itself cannot disentangle the impacts of the experimental intervention from those of the junk term. Only our knowledge of the structure of the comparison and the plausibility of rival hypotheses would permit us to attribute the statistically significant difference to the experimental intervention. In this sense the finding of a statistically significant difference between the two groups does not in itself signal the existence of an experimental impact. Only the further assumption that the junk terms have the same distribution permits such a conclusion. Likewise when there is no clear separation of the two sets of observations, and the null hypothesis is strongly supported, only knowledge beyond that incorporated in the significance test can rule out the possibility that a real experimental impact was offset by a bias in the comparison.

With this sense of the true role of the significance test, we take up the matter of actually doing the Mann-Whitney test. The test is based not on the actual values of the two sets of observations but on their pooled ranks. The two sets of data are merged and rank-ordered, with rank 1 given to the smallest value, rank 2 to the second smallest, and so forth. We will assume for purposes of discussion that the two sets of observations arise from study and control groups, with the understanding that the same procedures would apply if they arose from before-and-after, time-series, or Jack Sprat designs.

If the study and control scores arise from the same distribution, then there should be good mixing of ranks, with neither group concentrated at either high or low ranks. If there is a clear separation between the groups, with one group having predominantly low ranks and the other predominantly high ranks, then the null hypothesis that the two groups' scores share a common distribution becomes hard to believe.

The essential idea of the Mann-Whitney test is best seen in a small sample. Suppose there are 2 study and 3 control group cases for comparison. Each has a score or observed value. Using these values, we rank-order 5 cases, giving rank 1 to the case with the lowest score and rank 5 to the case with the highest (for now we assume there are no ties). There are $\binom{5}{2} = 10$ possible rankings, which are listed in table 11.5. The key idea is that under the null hypothesis study and control cases are samples from the same distribution; each of the 10 possible combinations has the same probability of occuring (that is, $1/10$).

This fact permits us to elaborate the steps common to all significance tests:

1. Choose a test statistic.
2. Find the distribution of the test statistic under the null hypothesis.

Table 11.5
Distribution of test statistic for Mann-Whitney test

Rank:	1	2	3	4	5	Test statistic = sum of ranks of smaller group	Probability of test statistic under null hypothesis
Group:	S	S	C	C	C	3	1/10
	S	C	S	C	C	4	1/10
	S	C	C	S	C	5	
	C	S	S	C	C	5	2/10
	S	C	C	C	S	6	
	C	S	C	S	C	6	2/10
	C	S	C	C	S	7	
	C	C	S	S	C	7	2/10
	C	C	S	C	S	8	1/10
	C	C	C	S	S	9	1/10
							Total 10/10

3. Determine the descriptive level of significance, which is the probability of encountering results as extreme as those observed if the null hypothesis be true.

We examine now in detail how these steps proceed for the Mann-Whitney test.

1. *Choose a test statistic.* We chose the sum of the ranks assigned to the group having the fewer cases. If this sum is small, the smaller group tends to have lower scores; if large, the smaller group tends to have higher scores. In the example with 2 study cases and 3 control cases we concentrate on the rank sum of the study group, since it is smaller (if the two groups have the same number of cases, pick either group). For samples of 2 and 3 cases, the rank sum of the smaller groups will be a number between 3 (the sum of the two lowest ranks) and 9 (the sum of the two highest ranks).

2. *Find the distribution of the test statistic.* We know that, if the null hypothesis is correct, each possible ordering of the cases is equally likely. We simply determine those arrangements which have the same rank sums and add their probabilities to find the probability of observing that particular rank sum if the null hypothesis is true. See table 11.5.

3. Determine the descriptive level of significance of the observed value of the test statistic. Suppose the results of the comparison are

Rank: 1   2   3   4   5
Group: S C C S C

The rank sum of the smaller group is $1 + 4 = 5$. Under the null hypothesis,

$$\text{Prob}\begin{bmatrix} \text{rank} \\ \text{sum} \geq 5 \end{bmatrix} \begin{matrix} \text{null} \\ \text{hypothesis} \end{matrix} = \text{Prob}\begin{bmatrix} \text{rank} \\ \text{sum} = 3 \end{bmatrix} \begin{matrix} \text{null} \\ \text{hypothesis} \end{matrix}$$

$$+ \text{Prob}\begin{bmatrix} \text{rank} \\ \text{sum} = 4 \end{bmatrix} \begin{matrix} \text{null} \\ \text{hypothesis} \end{matrix}$$

$$+ \text{Prob}\begin{bmatrix} \text{rank} \\ \text{sum} = 5 \end{bmatrix} \begin{matrix} \text{null} \\ \text{hypothesis} \end{matrix}$$

$$= \tfrac{1}{10} + \tfrac{1}{10} + \tfrac{2}{10} = \tfrac{4}{10}.$$

Thus if the null were true, we would expect—in repeated experiments comparing 2 study cases against 3 controls—that fully 4 in 10 of such comparisons would exhibit this degree of separation between the groups just because of sampling variations.

By traditional standards, the finding of 4 in 10 odds of seeing such extreme differences between the groups due solely to random sampling fluctuations would lead to dismissal of the differences as statistical flukes. In fact, it is traditional to dismiss differences that would occur only once in 20 trials, let alone 4 times in 10. However, you need not feel bound to this traditional rule if it does not suit the circumstances of your own situation. For instance, suppose that you are desperately searching for even a slightly less costly way to deliver some service. You may be more than willing to bankroll many wild goose chases, just so long as you do not overlook one promising innovation. In such a circumstance, the finding that differences as great or greater would arise "only" 4 times in 10 due to sampling fluctuations may lead you to reject the null hypothesis of no difference between study and control groups. On the other hand, if the last thing you want is to stake your reputation on another search for the unicorn, you may insist that differences between the groups be so extreme that only once in 1,000 trials could chance alone create such differences before you reject the null hypothesis and endorse the experimental treatment. Your reaction to the descriptive level of significance will always be largely subjective.

A practical matter remains to be discussed regarding step 3: determining the probability that the observed differences could have arisen by chance if the null hypothesis were true. In theory, we can list all the possible combinations of ranks for two groups. However, in practice the number of combinations can grow so large even for relatively small numbers of cases that no one could spend the time listing all the possibilities. If the smaller group has n cases and the larger group has m cases, then there are $\binom{n+m}{n}$ combinations. For example, in the earlier example evaluating incomes of those with and without exposure to a job-training program (table 11.3), there were $n = 12$ control workers and $m = 13$ study workers; therefore there are $\binom{25}{12}$ or over 5 million combinations. What we do in practice is use tables if the samples have a few cases in each group, say, 3 or 4 in the smaller group (see Mosteller and Rourke, table A-9), and use an approximation based on the Gaussian distribution for larger samples.

The rank sum of the smaller group is a random variable with a mean and standard deviation, and it will vary from trial to trial. Under the null hypothesis, the mean and standard deviation depend only on the sizes of the two groups. Letting n and m represent the number of cases in the smaller and larger groups, respectively, and letting μ and σ represent the mean and standard deviation of the smaller group's rank sum under the null hypothesis, then

$$\mu = \frac{n(n + m + 1)}{2}$$

$$\sigma = \sqrt{\frac{mn(n + m + 1)}{12}}.$$

For instance, in our job-training example we would expect under the null hypothesis that the rank sum for the smaller (control) group will hover around

$$\mu = \frac{(12)(12 + 13 + 1)}{2} = 156.0$$

with standard deviation

$$\sigma = \sqrt{\frac{(13)(12)(12 + 13 + 1)}{12}} = 18.38.$$

Thus any rank sum above 156.0 will be considered high and evidence of higher incomes in the control group.

Let us return to the job-training data and analyze them using the Mann-Whitney procedure. Recall that the randomized with-and-without design led to the unlucky conclusion that the control group had higher incomes when in fact the opposite was true. Will the Mann-Whitney test at least save us from drawing the wrong conclusion, even if we fail to reach the right one? When the data in table 11.3 are rank-ordered, the sum of the ranks assigned to control cases is 164, slightly higher than 156 but less than one standard deviation away.

Now the Gaussian approximation is as follows:

Prob [rank sum of controls $\geq$ 164 | null hypothesis true]

$$\approx \text{Prob}\left[Z \geq \frac{164 - \mu - 1/2}{\sigma}\right],$$

where Z is the standardized Gaussian variable and the $-1/2$ is a continuity correction that improves the approximation. In this case $\mu = 156$ and $\sigma = 18.38$, so

$$\frac{\text{Descriptive level}}{\text{of significance}} = \text{Prob [rank sum of controls} \geq 164 \,|\, \text{null true]}$$

$$\approx \text{Prob}\left[Z \geq \frac{164 - 156 - 1/2}{18.38}\right]$$

$$= \text{Prob}\,[Z \geq 0.41]$$

$$= 0.34.$$

Thus there is about one chance in three that, when the null is true, a rank sum for the controls could equal or exceed 164 just because of sampling fluctuations. Most planners would conclude from this that there is no real evidence of difference between the two groups: incomes in study and control groups appear to have the same distribution. Thus the conclusion of the comparison would be that the job-training program had no impact. This conclusion is wrong, but not as wrong as the conclusion that the job-training program actually hurt its students.

Some final notes on the Mann-Whitney test. First, the continuity correction becomes $+1/2$ instead of $-1/2$ if the test statistic is lower (rather than higher) than μ. Second, full details of the treatment of ties in ranking

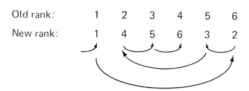

And again,

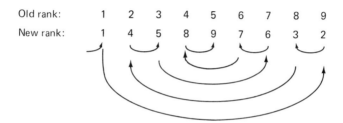

Figure 11.4
The method of assigning ranks when the Mann-Whitney test is used to compare the
dispersion of two data sets

are available in Mosteller and Rouke. Third, the same test can be used
to assess a different kind of experimental impact.

The Mann-Whitney procedure as described earlier (and the Bayesian
estimation procedure using Student's t) really acts as a test for a difference
in the central tendency of two sets of observations. A program, however,
may seek to influence the *dispersion* of a variable of interest rather than
its central tendency. An example would be a zoning policy aimed at
minimizing drastic fluctuations in population growth while permitting
the total population to rise. By modifying the assignment of ranks to
cases, the Mann-Whitney test can be applied to the assessment of changes
in dispersion. Once the cases have been rank-ordered as before, new
ranks can be assigned in a kind of leap-frog fashion as illustrated in
figure 11.4. Once these new ranks have been assigned, the Mann-Whitney
test procedes as before. The idea in this case is that, if the smaller group

has a small rank sum, it will generally have both the lowest and highest values and therefore have a greater variability around the common measure of central tendency. Likewise large rank sums indicate values clustered in the middle of the pooled distribution, meaning that the smaller group has less variability. A housing policy that seeks to increase the income mix of a neighborhood without shifting the mean income might be evaluated using the Mann-Whitney test this way in a before-and-after comparison.

Summary

Comparisons of two programs should be examined for both internal and external validity. A comparison has internal validity if the observed difference can properly be attributed to the experimental impact rather than to other influences, biased comparison, measurement artifact, or sampling accident. Rival hypotheses can be rendered implausible by carefully structuring the comparison and by using the techniques of statistical inference. A comparison has external validity if the results are generalizable to other settings. This generalizability requires that the experimental intervention be applied to representative cases behaving in representative ways and that it be possible to replicate the intervention. There is often a trade-off between the goals of internal and external validity, although greater investment in the experimental design usually enhances both.

The primitive demonstration project can be improved by investing either in pretest information (giving a before-and-after design) or in a control group (giving a with-and-without design). Adding both pretests and a control group leads to the Jack Sprat design, while making repeated observations before and after the intervention produces a time-series design. When working with control groups, matching pairs of cases helps somewhat to make the comparison fair, but random assignment of cases to groups offers strong statistical assurance that the two groups are equivalent and is definitely to be preferred to matching. Random assignment of matched pairs is better yet. The ability of a design to estimate accurately an experimental impact can be assessed with the Monte Carlo method, which uses repeated random trials to simulate the effects of various threats to internal validity.

Statistical inference has an important, but not primary, role in the comparison of alternative policies. The better the experimental design and the better the intervention, the less need there will be for reliance on statistical techniques to document the benefits of the innovation. Estimates of experimental impact and tests of significance presume that the design has been carefully chosen to minimize contamination of the data, so statistical inference alone cannot provide unambiguous conclusions about the superiority of one program over another. If the differences are approximately Gaussian, Bayesian estimation methods using Student's t distribution can be applied to differences from pretest to posttest, or between members of matched pairs, or both as in the Jack Sprat design.

If the data are at least ordinal in level of measurement, the Mann-Whitney procedure can be applied to test the null hypothesis that two groups of observations arose from the same distribution. The procedure begins by pooling the observations, ordering them, then assigning ranks. The assignment of ranks can be done in either of two ways, depending on whether the presumed experimental impact is a shift in central tendency or a shift in dispersion. When the data arise from a Jack Sprat design, the two sets of observations should be the differences from pretest to posttest in the study and control groups. When the data arise from a time-series design, the presence of trends usually requires that the two sets of observations be made up of differences between successive observations (or ratios of successive observations) from both the before and after periods. When the number of cases in each group is greater than about 3 or 4, the descriptive level of significance for the Mann-Whitney test can be readily approximated using the Gaussian distribution.

References and Readings

Brooks, R. "Bayesian Analysis of a Two-sample Problem Based on the Rank Order Statistic." *Journal of the Royal Statistical Society* (B) 40 (1978): 50–57

Campbell, D., and H. Ross. "The Connecticut Crackdown on Speeding: Time Series Data in Quasi-Experimental Analysis." *Law and Society Review* 3 (1968): 33–53.

Campbell, D., and J. Stanley, pp. 1–24, 34–43, 55–56.

Campbell, D. "Reforms as Experiments." *American Psychologist* 24 (1969): 409–429.

Conner, R. "Selecting a Control Group: An Analysis of the Randomization Process in Twelve Social Reform Programs." *Evaluation Quarterly* 1 (1977): 195–244.

Cook, T., and C. Reichardt. "Statistical Analysis of Nonequivalent Control Group Designs: A Guide to Some Current Literature." *Evaluation* 3 (1976): 136–138.

Gilbert, J. P., R. J. Light, and F. Mosteller. "Assessing Social Innovations: An Empirical Basis for Policy." In *Evaluation and Experiment*. Edited by C. A. Bennett and A. A. Lumsdaine. New York: Academic Press, 1975, pp. 39–182.

Gould, E. "Park Areas and Open Spaces in American and European Cities." *Publications of the American Statistical Association* 1 (1888): 49–61.

Kalton, G., M. Collins, and L. Brook. "Experiments in Wording Opinion Questions." *Applied Statistics* 27 (1978): 149–161.

Mosteller and Rourke. "Ranking Methods for Two Independent Samples," chapter 3, "Normal Approximation for the Mann-Whitney Test," chapter 4, and "Wilcoxon's Signed Rank Test," chapter 5, pp. 54–72, 73–88, 89–102.

Roos, N. "Contrasting Social Experimentation with Retrospective Evaluation." *Public Policy* 23 (1975): 241–257.

Wilson, J. "Social Experimentation and Public Policy." *Public Policy* 22 (1974): 15–37.

Problems

11.1

To alleviate backlogs in its emergency department, Carney Hospital in Boston began in late July 1977 to staff the department with two physicians rather than one during the hours 4 to 9 P.M. Since the physician(s) at Carney Hospital must care for several patients simultaneously by rotating among several treatment rooms, it was expected that adding the second physician would reduce the inefficiencies of constantly switching attention from one patient to the next, thereby reducing the total time required to treat each patient. Since the second physician was hired on a trial basis, the hospital desired a quick evaluation of the impact of the second physician. The hospital's management information system provided data on treatment times. Willemain et al. obtained the following estimates of mean treatment times for patients whose treatment began during the indicated hours on the indicated days. Why do you think the dates and hours shown were chosen for analysis? Analyze the data to assess the impact of adding the second physician. Comment on your results.

Date patients treated	Hours during which treatment initiated	Number of physicians	Number of patients	Treatment time (minutes)		
				Mean	Standard deviation	Median
7/11/77	12–2 P.M.	1	10	125	92	98
7/11/77	4–7	1	16	34	27	30
7/12/77	12–2	1	13	99	69	78
7/12/77	4–7	1	21	76	58	49
7/13/77	12–2	1	8	55	38	56
7/13/77	4–7	1	16	69	74	49
7/16/77	12–2	1	9	38	31	25
7/16/77	4–7	1	9	72	68	48
7/17/77	12–2	1	13	44	45	31
7/17/77	4–7	1	15	58	49	48
7/25/77	12–2	1	12	99	65	82
7/25/77	4–7	2	12	88	89	75
7/26/77	12–2	1	11	76	31	82
7/26/77	4–7	2	11	68	62	48
7/27/77	12–2	1	11	49	43	28
7/27/77	4–7	2	15	102	79	83

(continued)

Date patients treated	Hours during which treatment initiated	Number of physicians	Number of patients	Treatment time (minutes)		
				Mean	Standard deviation	Median
7/30/77	12–2	1	8	78	48	86
7/30/77	4–7	2	14	62	35	49
7/31/77	12–2	1	11	62	37	58
7/31/77	4–7	2	17	80	57	67

11.2

Use the data in problem 11.1 to test whether the relative variability of treatment times is different during the periods 12 to 2 P.M. and 4 to 7 P.M. when there is only one physician working.

11.3

The following data show the number of mobile home shipments to dealers and the estimated volume of retail sales in each year from 1952 to 1976. What can you say using these data about (a) the introduction of 10-foot wide mobile homes into mass production in 1955, (b) the introduction of 12-foot wide homes in 1962, (c) the methods used to estimate shipments and retail sales?

Mobile home shipments to dealers in the United States and estimated retail sales, 1952 to 1976[a]

Year	Mobile home shipments	Retail sales (estimated in thousands of dollars)
1952	83,000	$320,000
1953	76,900	322,000
1954	76,000	325,000
1955	111,900	462,000
1956	124,330	622,000
1957	119,300	596,000
1958	102,000	510,000
1959	120,500	602,000
1960	103,700	518,000
1961	90,200	505,000
1962	118,000	661,000

Mobile home shipments to dealers (continued)

Year	Mobile home shipments	Retail sales (estimated in thousands of dollars)
1963	150,840	862,064
1964	191,320	1,071,392
1965	216,470	1,212,232
1966	217,300	1,238,610
1967	240,360	1,370,052
1968	317,950	1,907,700
1969	412,690	2,496,775
1970	401,190	2,451,271
1971	496,570	3,297,225
1972	575,940	4,153,103
1973	566,920	4,406,382
1974	329,300	3,213,681
1975	212,690	2,432,661
1976	249,870	3,184,427

[a]Prior to 1951, production varied from 1,300 in 1930 upward to 83,000 in 1952. Ten-foot wide homes came into mass production in 1955; 12-foot wide homes came into mass production in 1962.

Appendix A: Table of the Binomial Distribution

Binomial distribution $\binom{N}{n} P^n (1 - P)^{N-n}$

						P					
N	n	0.05	0.10	0.15	0.20	0.25	0.30	0.35	0.40	0.45	0.50
1	0	0.9500	0.9000	0.8500	0.8000	0.7500	0.7000	0.6500	0.6000	0.5500	0.5000
	1	0.0500	0.1000	0.1500	0.2000	0.2500	0.3000	0.3500	0.4000	0.4500	0.5000
2	0	0.9025	0.8100	0.7225	0.6400	0.5625	0.4900	0.4225	0.3600	0.3025	0.2500
	1	0.0950	0.1800	0.2550	0.3200	0.3750	0.4200	0.4550	0.4800	0.4950	0.5000
	2	0.0025	0.0100	0.0225	0.0400	0.0625	0.0900	0.1225	0.1600	0.2025	0.2500
3	0	0.8574	0.7290	0.6141	0.5120	0.4219	0.3430	0.2746	0.2160	0.1664	0.1250
	1	0.1354	0.2430	0.3251	0.3840	0.4219	0.4410	0.4436	0.4320	0.4084	0.3750
	2	0.0071	0.0270	0.0574	0.0960	0.1406	0.1890	0.2389	0.2880	0.3341	0.3750
	3	0.0001	0.0010	0.0034	0.0080	0.0156	0.0270	0.0429	0.0640	0.0911	0.1250
4	0	0.8145	0.6561	0.5220	0.4096	0.3164	0.2401	0.1785	0.1296	0.0915	0.0625
	1	0.1715	0.2916	0.3685	0.4096	0.4219	0.4116	0.3845	0.3456	0.2995	0.2500
	2	0.0135	0.0486	0.0975	0.1536	0.2109	0.2646	0.3105	0.3456	0.3675	0.3750
	3	0.0005	0.0036	0.0115	0.0256	0.0469	0.0756	0.1115	0.1536	0.2005	0.2500
	4	0.0000	0.0001	0.0005	0.0016	0.0039	0.0081	0.0150	0.0256	0.0410	0.0625
5	0	0.7738	0.5905	0.4437	0.3277	0.2373	0.1681	0.1160	0.0778	0.0503	0.0312
	1	0.2036	0.3280	0.3915	0.4096	0.3955	0.3602	0.3124	0.2592	0.2059	0.1562
	2	0.0214	0.0729	0.1382	0.2048	0.2637	0.3087	0.3364	0.3456	0.3369	0.3125
	3	0.0011	0.0081	0.0244	0.0512	0.0879	0.1323	0.1811	0.2304	0.2757	0.3125
	4	0.0000	0.0004	0.0022	0.0064	0.0146	0.0284	0.0488	0.0768	0.1128	0.1562
	5	0.0000	0.0000	0.0001	0.0003	0.0010	0.0024	0.0053	0.0102	0.0185	0.0312
6	0	0.7351	0.5314	0.3771	0.2621	0.1780	0.1176	0.0754	0.0467	0.0277	0.0156
	1	0.2321	0.3543	0.3993	0.3932	0.3560	0.3025	0.2437	0.1866	0.1359	0.0938
	2	0.0305	0.0984	0.1762	0.2458	0.2966	0.3241	0.3280	0.3110	0.2780	0.2344
	3	0.0021	0.0146	0.0415	0.0819	0.1318	0.1852	0.2355	0.2765	0.3032	0.3125
	4	0.0001	0.0012	0.0055	0.0154	0.0330	0.0595	0.0951	0.1382	0.1861	0.2344
	5	0.0000	0.0001	0.0004	0.0015	0.0044	0.0102	0.0205	0.0369	0.0609	0.0938
	6	0.0000	0.0000	0.0000	0.0001	0.0002	0.0007	0.0018	0.0041	0.0083	0.0156

Source: *Handbook of Tables for Probability and Statistics*, 2d ed., W. H. Beyer, ed. (West Palm Beach, Fla.: CRC Press, Inc., 1968). Adapted with permission of the publisher.
Note: Linear interpolations with respect to P will in general be accurate at most to two decimal places. For values of $P \geq 0.50$ use $\binom{N}{n} P^n (1 - P)^{N-n} = \binom{N}{N-n}(1 - P)^n P^{N-n}$. For instance, for $P = 0.55$

$$\binom{5}{3}(0.55)^3(0.45)^2 = \binom{5}{2}(0.45)^3(0.55)^2 = 0.3369.$$

Binomial distribution $\binom{N}{n}P^n(1-P)^{N-n}$

N	n	0.05	0.10	0.15	0.20	P 0.25	0.30	0.35	0.40	0.45	0.50
7	0	0.6983	0.4783	0.3206	0.2097	0.1335	0.0824	0.0490	0.0280	0.0152	0.0078
	1	0.2573	0.3720	0.3960	0.3670	0.3115	0.2471	0.1848	0.1306	0.0872	0.0547
	2	0.0406	0.1240	0.2097	0.2753	0.3115	0.3177	0.2985	0.2613	0.2140	0.1641
	3	0.0036	0.0230	0.0617	0.1147	0.1730	0.2269	0.2679	0.2903	0.2918	0.2734
	4	0.0002	0.0026	0.0109	0.0287	0.0577	0.0972	0.1442	0.1935	0.2388	0.2734
	5	0.0000	0.0002	0.0012	0.0043	0.0115	0.0250	0.0466	0.0774	0.1172	0.1641
	6	0.0000	0.0000	0.0001	0.0004	0.0013	0.0036	0.0084	0.0172	0.0320	0.0547
	7	0.0000	0.0000	0.0000	0.0000	0.0001	0.0002	0.0006	0.0016	0.0037	0.0078
8	0	0.6634	0.4305	0.2725	0.1678	0.1001	0.0576	0.0319	0.0168	0.0084	0.0039
	1	0.2793	0.3826	0.3847	0.3355	0.2670	0.1977	0.1373	0.0896	0.0548	0.0312
	2	0.0515	0.1488	0.2376	0.2936	0.3115	0.2965	0.2587	0.2090	0.1569	0.1094
	3	0.0054	0.0331	0.0839	0.1468	0.2076	0.2541	0.2786	0.2787	0.2568	0.2188
	4	0.0004	0.0046	0.0185	0.0459	0.0865	0.1361	0.1875	0.2322	0.2627	0.2734
	5	0.0000	0.0004	0.0026	0.0092	0.0231	0.0467	0.0808	0.1239	0.1719	0.2188
	6	0.0000	0.0000	0.0002	0.0011	0.0038	0.0100	0.0217	0.0413	0.0703	0.1094
	7	0.0000	0.0000	0.0000	0.0001	0.0004	0.0012	0.0033	0.0079	0.0164	0.0312
	8	0.0000	0.0000	0.0000	0.0000	0.0000	0.0001	0.0002	0.0007	0.0017	0.0039
9	0	0.6302	0.3874	0.2316	0.1342	0.0751	0.0404	0.0207	0.0101	0.0046	0.0020
	1	0.2985	0.3874	0.3679	0.3020	0.2253	0.1556	0.1004	0.0605	0.0339	0.0176
	2	0.0629	0.1722	0.2597	0.3020	0.3003	0.2668	0.2162	0.1612	0.1110	0.0703
	3	0.0077	0.0446	0.1069	0.1762	0.2336	0.2668	0.2716	0.2508	0.2119	0.1641
	4	0.0006	0.0074	0.0283	0.0661	0.1168	0.1715	0.2194	0.2508	0.2600	0.2461
	5	0.0000	0.0008	0.0050	0.0165	0.0389	0.0735	0.1181	0.1672	0.2128	0.2461
	6	0.0000	0.0001	0.0006	0.0028	0.0087	0.0210	0.0424	0.0743	0.1160	0.1641
	7	0.0000	0.0000	0.0000	0.0003	0.0012	0.0039	0.0098	0.0212	0.0407	0.0703
	8	0.0000	0.0000	0.0000	0.0000	0.0001	0.0004	0.0013	0.0035	0.0083	0.0176
	9	0.0000	0.0000	0.0000	0.0000	0.0000	0.0000	0.0001	0.0003	0.0008	0.0020
10	0	0.5987	0.3487	0.1969	0.1074	0.0563	0.0282	0.0135	0.0060	0.0025	0.0010
	1	0.3151	0.3874	0.3474	0.2684	0.1877	0.1211	0.0725	0.0403	0.0207	0.0098
	2	0.0746	0.1937	0.2759	0.3020	0.2816	0.2335	0.1757	0.1209	0.0763	0.0439
	3	0.0105	0.0574	0.1298	0.2013	0.2503	0.2668	0.2522	0.2150	0.1665	0.1172
	4	0.0010	0.0112	0.0401	0.0881	0.1460	0.2001	0.2377	0.2508	0.2384	0.2051
	5	0.0001	0.0015	0.0085	0.0264	0.0584	0.1029	0.1536	0.2007	0.2340	0.2461
	6	0.0000	0.0001	0.0012	0.0055	0.0162	0.0368	0.0689	0.1115	0.1596	0.2051
	7	0.0000	0.0000	0.0001	0.0008	0.0031	0.0090	0.0212	0.0425	0.0746	0.1172
	8	0.0000	0.0000	0.0000	0.0001	0.0004	0.0014	0.0043	0.0106	0.0229	0.0439
	9	0.0000	0.0000	0.0000	0.0000	0.0000	0.0001	0.0005	0.0016	0.0042	0.0098
	10	0.0000	0.0000	0.0000	0.0000	0.0000	0.0000	0.0000	0.0001	0.0003	0.0010

Binomial distribution $\binom{N}{n}P^n(1 - P)^{N-n}$

						P					
N	n	0.05	0.10	0.15	0.20	0.25	0.30	0.35	0.40	0.45	0.50
11	0	0.5688	0.3138	0.1673	0.0859	0.0422	0.0198	0.0088	0.0036	0.0014	0.0004
	1	0.3293	0.3835	0.3248	0.2362	0.1549	0.0932	0.0518	0.0266	0.0125	0.0055
	2	0.0867	0.2131	0.2866	0.2953	0.2581	0.1998	0.1395	0.0887	0.0513	0.0269
	3	0.0137	0.0710	0.1517	0.2215	0.2581	0.2568	0.2254	0.1774	0.1259	0.0806
	4	0.0014	0.0158	0.0536	0.1107	0.1721	0.2201	0.2428	0.2365	0.2060	0.1611
	5	0.0001	0.0025	0.0132	0.0388	0.0803	0.1321	0.1830	0.2207	0.2360	0.2256
	6	0.0000	0.0003	0.0023	0.0097	0.0268	0.0566	0.0985	0.1471	0.1931	0.2256
	7	0.0000	0.0000	0.0003	0.0017	0.0064	0.0173	0.0379	0.0701	0.1128	0.1611
	8	0.0000	0.0000	0.0000	0.0002	0.0011	0.0037	0.0102	0.0234	0.0462	0.0806
	9	0.0000	0.0000	0.0000	0.0000	0.0001	0.0005	0.0018	0.0052	0.0126	0.0269
	10	0.0000	0.0000	0.0000	0.0000	0.0000	0.0000	0.0002	0.0007	0.0021	0.0054
	11	0.0000	0.0000	0.0000	0.0000	0.0000	0.0000	0.0000	0.0000	0.0002	0.0005
12	0	0.5404	0.2824	0.1422	0.0687	0.0317	0.0138	0.0057	0.0022	0.0008	0.0002
	1	0.3413	0.3766	0.3012	0.2062	0.1267	0.0712	0.0368	0.0174	0.0075	0.0029
	2	0.0988	0.2301	0.2924	0.2835	0.2323	0.1678	0.1088	0.0639	0.0339	0.0161
	3	0.0173	0.0852	0.1720	0.2362	0.2581	0.2397	0.1954	0.1419	0.0923	0.0537
	4	0.0021	0.0213	0.0683	0.1329	0.1936	0.2311	0.2367	0.2128	0.1700	0.1208
	5	0.0002	0.0038	0.0193	0.0532	0.1032	0.1585	0.2039	0.2270	0.2225	0.1934
	6	0.0000	0.0005	0.0040	0.0155	0.0401	0.0792	0.1281	0.1766	0.2124	0.2256
	7	0.0000	0.0000	0.0006	0.0033	0.0115	0.0291	0.0591	0.1009	0.1489	0.1934
	8	0.0000	0.0000	0.0001	0.0005	0.0024	0.0078	0.0199	0.0420	0.0762	0.1208
	9	0.0000	0.0000	0.0000	0.0001	0.0004	0.0015	0.0048	0.0125	0.0277	0.0537
	10	0.0000	0.0000	0.0000	0.0000	0.0000	0.0002	0.0008	0.0025	0.0068	0.0161
	11	0.0000	0.0000	0.0000	0.0000	0.0000	0.0000	0.0001	0.0003	0.0010	0.0029
	12	0.0000	0.0000	0.0000	0.0000	0.0000	0.0000	0.0000	0.0000	0.0001	0.0002
13	0	0.5133	0.2542	0.1209	0.0550	0.0238	0.0097	0.0037	0.0013	0.0004	0.0001
	1	0.3512	0.3672	0.2774	0.1787	0.1029	0.0540	0.0259	0.0113	0.0045	0.0016
	2	0.1109	0.2448	0.2937	0.2680	0.2059	0.1388	0.0836	0.0453	0.0220	0.0095
	3	0.0214	0.0997	0.1900	0.2457	0.2517	0.2181	0.1651	0.1107	0.0660	0.0349
	4	0.0028	0.0277	0.0838	0.1535	0.2097	0.2337	0.2222	0.1845	0.1350	0.0873
	5	0.0003	0.0055	0.0266	0.0691	0.1258	0.1803	0.2154	0.2214	0.1989	0.1571
	6	0.0000	0.0008	0.0063	0.0230	0.0559	0.1030	0.1546	0.1968	0.2169	0.2095
	7	0.0000	0.0001	0.0011	0.0058	0.0186	0.0442	0.0833	0.1312	0.1775	0.2095
	8	0.0000	0.0000	0.0001	0.0011	0.0047	0.0142	0.0336	0.0656	0.1089	0.1571
	9	0.0000	0.0000	0.0000	0.0001	0.0009	0.0034	0.0101	0.0243	0.0495	0.0873
	10	0.0000	0.0000	0.0000	0.0000	0.0001	0.0006	0.0022	0.0065	0.0162	0.0349
	11	0.0000	0.0000	0.0000	0.0000	0.0000	0.0001	0.0003	0.0012	0.0036	0.0095
	12	0.0000	0.0000	0.0000	0.0000	0.0000	0.0000	0.0000	0.0001	0.0005	0.0016
	13	0.0000	0.0000	0.0000	0.0000	0.0000	0.0000	0.0000	0.0000	0.0000	0.0001

Binomial distribution $\binom{N}{n}P^n(1-P)^{N-n}$

N	n	0.05	0.10	0.15	0.20	P. 0.25	0.30	0.35	0.40	0.45	0.50
14	0	0.4877	0.2288	0.1028	0.0440	0.0178	0.0068	0.0024	0.0008	0.0002	0.0001
	1	0.3593	0.3559	0.2539	0.1539	0.0832	0.0407	0.0181	0.0073	0.0027	0.0009
	2	0.1229	0.2570	0.2912	0.2501	0.1802	0.1134	0.0634	0.0317	0.0141	0.0056
	3	0.0259	0.1142	0.2056	0.2501	0.2402	0.1943	0.1366	0.0845	0.0462	0.0222
	4	0.0037	0.0349	0.0998	0.1720	0.2202	0.2290	0.2022	0.1549	0.1040	0.0611
	5	0.0004	0.0078	0.0352	0.0860	0.1468	0.1963	0.2178	0.2066	0.1701	0.1222
	6	0.0000	0.0013	0.0093	0.0322	0.0734	0.1262	0.1759	0.2066	0.2088	0.1833
	7	0.0000	0.0002	0.0019	0.0092	0.0280	0.0618	0.1082	0.1574	0.1952	0.2095
	8	0.0000	0.0000	0.0003	0.0020	0.0082	0.0232	0.0510	0.0918	0.1398	0.1833
	9	0.0000	0.0000	0.0000	0.0003	0.0018	0.0066	0.0183	0.0408	0.0762	0.1222
	10	0.0000	0.0000	0.0000	0.0000	0.0003	0.0014	0.0049	0.0136	0.0312	0.0611
	11	0.0000	0.0000	0.0000	0.0000	0.0000	0.0002	0.0010	0.0033	0.0093	0.0222
	12	0.0000	0.0000	0.0000	0.0000	0.0000	0.0000	0.0001	0.0005	0.0019	0.0056
	13	0.0000	0.0000	0.0000	0.0000	0.0000	0.0000	0.0000	0.0001	0.0002	0.0009
	14	0.0000	0.0000	0.0000	0.0000	0.0000	0.0000	0.0000	0.0000	0.0000	0.0001
15	0	0.4633	0.2059	0.0874	0.0352	0.0134	0.0047	0.0016	0.0005	0.0001	0.0000
	1	0.3658	0.3432	0.2312	0.1319	0.0668	0.0305	0.0126	0.0047	0.0016	0.0005
	2	0.1348	0.2669	0.2856	0.2309	0.1559	0.0916	0.0476	0.0219	0.0090	0.0032
	3	0.0307	0.1285	0.2184	0.2501	0.2252	0.1700	0.1110	0.0634	0.0318	0.0139
	4	0.0049	0.0428	0.1156	0.1876	0.2252	0.2186	0.1792	0.1268	0.0780	0.0417
	5	0.0006	0.0105	0.0449	0.1032	0.1651	0.2061	0.2123	0.1859	0.1404	0.0916
	6	0.0000	0.0019	0.0132	0.0430	0.0917	0.1472	0.1906	0.2066	0.1914	0.1527
	7	0.0000	0.0003	0.0030	0.0138	0.0393	0.0811	0.1319	0.1771	0.2013	0.1964
	8	0.0000	0.0000	0.0005	0.0035	0.0131	0.0348	0.0710	0.1181	0.1647	0.1964
	9	0.0000	0.0000	0.0001	0.0007	0.0034	0.0116	0.0298	0.0612	0.1048	0.1527
	10	0.0000	0.0000	0.0000	0.0001	0.0007	0.0030	0.0096	0.0245	0.0515	0.0916
	11	0.0000	0.0000	0.0000	0.0000	0.0001	0.0006	0.0024	0.0074	0.0191	0.0417
	12	0.0000	0.0000	0.0000	0.0000	0.0000	0.0001	0.0004	0.0016	0.0052	0.0139
	13	0.0000	0.0000	0.0000	0.0000	0.0000	0.0000	0.0001	0.0003	0.0010	0.0032
	14	0.0000	0.0000	0.0000	0.0000	0.0000	0.0000	0.0000	0.0000	0.0001	0.0005
	15	0.0000	0.0000	0.0000	0.0000	0.0000	0.0000	0.0000	0.0000	0.0000	0.0000

Binomial distribution $\binom{N}{n}P^n(1 - P)^{N-n}$

						P					
N	n	0.05	0.10	0.15	0.20	0.25	0.30	0.35	0.40	0.45	0.50
16	0	0.4401	0.1853	0.0743	0.0281	0.0100	0.0033	0.0010	0.0003	0.0001	0.0000
	1	0.3706	0.3294	0.2097	0.1126	0.0535	0.0228	0.0087	0.0030	0.0009	0.0002
	2	0.1463	0.2745	0.2775	0.2111	0.1336	0.0732	0.0353	0.0150	0.0056	0.0018
	3	0.0359	0.1423	0.2285	0.2463	0.2079	0.1465	0.0888	0.0468	0.0215	0.0085
	4	0.0061	0.0514	0.1311	0.2001	0.2252	0.2040	0.1553	0.1014	0.0572	0.0278
	5	0.0008	0.0137	0.0555	0.1201	0.1802	0.2099	0.2008	0.1623	0.1123	0.0667
	6	0.0001	0.0028	0.0180	0.0550	0.1101	0.1649	0.1982	0.1983	0.1684	0.1222
	7	0.0000	0.0004	0.0045	0.0197	0.0524	0.1010	0.1524	0.1889	0.1969	0.1746
	8	0.0000	0.0001	0.0009	0.0055	0.0197	0.0487	0.0923	0.1417	0.1812	0.1964
	9	0.0000	0.0000	0.0001	0.0012	0.0058	0.0185	0.0442	0.0840	0.1318	0.1746
	10	0.0000	0.0000	0.0000	0.0002	0.0014	0.0056	0.0167	0.0392	0.0755	0.1222
	11	0.0000	0.0000	0.0000	0.0000	0.0002	0.0013	0.0049	0.0142	0.0337	0.0667
	12	0.0000	0.0000	0.0000	0.0000	0.0000	0.0002	0.0011	0.0040	0.0115	0.0278
	13	0.0000	0.0000	0.0000	0.0000	0.0000	0.0000	0.0002	0.0008	0.0029	0.0085
	14	0.0000	0.0000	0.0000	0.0000	0.0000	0.0000	0.0000	0.0001	0.0005	0.0018
	15	0.0000	0.0000	0.0000	0.0000	0.0000	0.0000	0.0000	0.0000	0.0001	0.0002
	16	0.0000	0.0000	0.0000	0.0000	0.0000	0.0000	0.0000	0.0000	0.0000	0.0000
17	0	0.4181	0.1668	0.0631	0.0225	0.0075	0.0023	0.0007	0.0002	0.0000	0.0000
	1	0.3741	0.3150	0.1893	0.0957	0.0426	0.0169	0.0060	0.0019	0.0005	0.0001
	2	0.1575	0.2800	0.2673	0.1914	0.1136	0.0581	0.0260	0.0102	0.0035	0.0010
	3	0.0415	0.1556	0.2359	0.2393	0.1893	0.1245	0.0701	0.0341	0.0144	0.0052
	4	0.9076	0.0605	0.1457	0.2093	0.2209	0.1868	0.1320	0.0796	0.0411	0.0182
	5	0.0010	0.0175	0.0668	0.1361	0.1914	0.2081	0.1849	0.1379	0.0875	0.0472
	6	0.0001	0.0039	0.0236	0.0680	0.1276	0.1784	0.1991	0.1839	0.1432	0.0944
	7	0.0000	0.0007	0.0065	0.0267	0.0668	0.1201	0.1685	0.1927	0.1841	0.1484
	8	0.0000	0.0001	0.0014	0.0084	0.0279	0.0644	0.1134	0.1606	0.1883	0.1855
	9	0.0000	0.0000	0.0003	0.0021	0.0093	0.0276	0.0611	0.1070	0.1540	0.1855
	10	0.0000	0.0000	0.0000	0.0004	0.0025	0.0095	0.0263	0.0571	0.1008	0.1484
	11	0.0000	0.0000	0.0000	0.0001	0.0005	0.0026	0.0090	0.0242	0.0525	0.0944
	12	0.0000	0.0000	0.0000	0.0000	0.0001	0.0006	0.0024	0.0081	0.0215	0.0472
	13	0.0000	0.0000	0.0000	0.0000	0.0000	0.0001	0.0005	0.0021	0.0068	0.0182
	14	0.0000	0.0000	0.0000	0.0000	0.0000	0.0000	0.0001	0.0004	0.0016	0.0052
	15	0.0000	0.0000	0.0000	0.0000	0.0000	0.0000	0.0000	0.0001	0.0003	0.0010
	16	0.0000	0.0000	0.0000	0.0000	0.0000	0.0000	0.0000	0.0000	0.0000	0.0001
	17	0.0000	0.0000	0.0000	0.0000	0.0000	0.0000	0.0000	0.0000	0.0000	0.0000

Binomial distribution $\binom{N}{n}P^n(1 - P)^{N-n}$

N	n	0.05	0.10	0.15	0.20	P 0.25	0.30	0.35	0.40	0.45	0.50
18	0	0.3972	0.1501	0.0536	0.0180	0.0056	0.0016	0.0004	0.0001	0.0000	0.0000
	1	0.3763	0.3002	0.1704	0.0811	0.0338	0.0126	0.0042	0.0012	0.0003	0.0001
	2	0.1683	0.2835	0.2556	0.1723	0.0958	0.0458	0.0190	0.0069	0.0022	0.0006
	3	0.0473	0.1680	0.2406	0.2297	0.1704	0.1046	0.0547	0.0246	0.0095	0.0031
	4	0.0093	0.0700	0.1592	0.2153	0.2130	0.1681	0.1104	0.0614	0.0291	0.0117
	5	0.0014	0.0218	0.0787	0.1507	0.1988	0.2017	0.1664	0.1146	0.0666	0.0327
	6	0.0002	0.0052	0.0301	0.0816	0.1436	0.1873	0.1941	0.1655	0.1181	0.0708
	7	0.0000	0.0010	0.0091	0.0350	0.0820	0.1376	0.1792	0.1892	0.1657	0.1214
	8	0.0000	0.0002	0.0022	0.0120	0.0376	0.0811	0.1327	0.1734	0.1864	0.1669
	9	0.0000	0.0000	0.0004	0.0033	0.0139	0.0386	0.0794	0.1284	0.1694	0.1855
	10	0.0000	0.0000	0.0001	0.0008	0.0042	0.0149	0.0385	0.0771	0.1248	0.1669
	11	0.0000	0.0000	0.0000	0.0001	0.0010	0.0046	0.0151	0.0374	0.0742	0.1214
	12	0.0000	0.0000	0.0000	0.0000	0.0002	0.0012	0.0047	0.0145	0.0354	0.0708
	13	0.0000	0.0000	0.0000	0.0000	0.0000	0.0002	0.0012	0.0045	0.0134	0.0327
	14	0.0000	0.0000	0.0000	0.0000	0.0000	0.0000	0.0002	0.0011	0.0039	0.0117
	15	0.0000	0.0000	0.0000	0.0000	0.0000	0.0000	0.0000	0.0002	0.0009	0.0031
	16	0.0000	0.0000	0.0000	0.0000	0.0000	0.0000	0.0000	0.0000	0.0001	0.0006
	17	0.0000	0.0000	0.0000	0.0000	0.0000	0.0000	0.0000	0.0000	0.0000	0.0001
	18	0.0000	0.0000	0.0000	0.0000	0.0000	0.0000	0.0000	0.0000	0.0000	0.0000
19	0	0.3774	0.1351	0.0456	0.0144	0.0042	0.0011	0.0003	0.0001	0.0000	0.0000
	1	0.3774	0.2852	0.1529	0.0685	0.0268	0.0093	0.0029	0.0008	0.0002	0.0000
	2	0.1787	0.2852	0.2428	0.1540	0.0803	0.0358	0.0138	0.0046	0.0013	0.0003
	3	0.0533	0.1796	0.2428	0.2182	0.1517	0.0869	0.0422	0.0175	0.0062	0.0018
	4	0.0112	0.0798	0.1714	0.2182	0.2023	0.1491	0.0909	0.0467	0.0203	0.0074
	5	0.0018	0.0266	0.0907	0.1636	0.2023	0.1916	0.1468	0.0933	0.0497	0.0222
	6	0.0002	0.0069	0.0374	0.0955	0.1574	0.1916	0.1844	0.1451	0.0949	0.0518
	7	0.0000	0.0014	0.0122	0.0443	0.0974	0.1525	0.1844	0.1797	0.1443	0.0961
	8	0.0000	0.0002	0.0032	0.0166	0.0487	0.0981	0.1489	0.1797	0.1771	0.1442
	9	0.0000	0.0000	0.0007	0.0051	0.0198	0.0514	0.0980	0.1464	0.1771	0.1762
	10	0.0000	0.0000	0.0001	0.0013	0.0066	0.0220	0.0528	0.0976	0.1449	0.1762
	11	0.0000	0.0000	0.0000	0.0003	0.0018	0.0077	0.0233	0.0532	0.0970	0.1442
	12	0.0000	0.0000	0.0000	0.0000	0.0004	0.0022	0.0083	0.0237	0.0529	0.0961
	13	0.0000	0.0000	0.0000	0.0000	0.0001	0.0005	0.0024	0.0085	0.0233	0.0518
	14	0.0000	0.0000	0.0000	0.0000	0.0000	0.0001	0.0006	0.0024	0.0082	0.0222
	15	0.0000	0.0000	0.0000	0.0000	0.0000	0.0000	0.0001	0.0005	0.0022	0.0074
	16	0.0000	0.0000	0.0000	0.0000	0.0000	0.0000	0.0000	0.0001	0.0005	0.0018
	17	0.0000	0.0000	0.0000	0.0000	0.0000	0.0000	0.0000	0.0000	0.0001	0.0003
	18	0.0000	0.0000	0.0000	0.0000	0.0000	0.0000	0.0000	0.0000	0.0000	0.0000
	19	0.0000	0.0000	0.0000	0.0000	0.0000	0.0000	0.0000	0.0000	0.0000	0.0000

Binomial distribution $\binom{N}{n} P^n (1 - P)^{N-n}$

N	n	0.05	0.10	0.15	0.20	P 0.25	0.30	0.35	0.40	0.45	0.50
20	0	0.3585	0.1216	0.0388	0.0115	0.0032	0.0008	0.0002	0.0000	0.0000	0.0000
	1	0.3774	0.2702	0.1368	0.0576	0.0211	0.0068	0.0020	0.0005	0.0001	0.0000
	2	0.1887	0.2852	0.2293	0.1369	0.0669	0.0278	0.0100	0.0031	0.0008	0.0002
	3	0.0596	0.1901	0.2428	0.2054	0.1339	0.0716	0.0323	0.0123	0.0040	0.0011
	4	0.0133	0.0898	0.1821	0.2182	0.1897	0.1304	0.0738	0.0350	0.0139	0.0046
	5	0.0022	0.0319	0.1028	0.1746	0.2023	0.1789	0.1272	0.0746	0.0365	0.0148
	6	0.0003	0.0089	0.0454	0.1091	0.1686	0.1916	0.1712	0.1244	0.0746	0.0370
	7	0.0000	0.0020	0.0160	0.0545	0.1124	0.1643	0.1844	0.1659	0.1221	0.0739
	8	0.0000	0.0004	0.0046	0.0222	0.0609	0.1144	0.1614	0.1797	0.1623	0.1201
	9	0.0000	0.0001	0.0011	0.0074	0.0271	0.0654	0.1158	0.1597	0.1771	0.1602
	10	0.0000	0.0000	0.0002	0.0020	0.0099	0.0308	0.0686	0.1171	0.1593	0.1762
	11	0.0000	0.0000	0.0000	0.0005	0.0030	0.0120	0.0336	0.0710	0.1185	0.1602
	12	0.0000	0.0000	0.0000	0.0001	0.0008	0.0039	0.0136	0.0355	0.0727	0.1201
	13	0.0000	0.0000	0.0000	0.0000	0.0002	0.0010	0.0045	0.0146	0.0366	0.0739
	14	0.0000	0.0000	0.0000	0.0000	0.0000	0.0002	0.0012	0.0049	0.0150	0.0370
	15	0.0000	0.0000	0.0000	0.0000	0.0000	0.0000	0.0003	0.0013	0.0049	0.0148
	16	0.0000	0.0000	0.0000	0.0000	0.0000	0.0000	0.0000	0.0003	0.0013	0.0046
	17	0.0000	0.0000	0.0000	0.0000	0.0000	0.0000	0.0000	0.0000	0.0002	0.0011
	18	0.0000	0.0000	0.0000	0.0000	0.0000	0.0000	0.0000	0.0000	0.0000	0.0002
	19	0.0000	0.0000	0.0000	0.0000	0.0000	0.0000	0.0000	0.0000	0.0000	0.0000
	20	0.0000	0.0000	0.0000	0.0000	0.0000	0.0000	0.0000	0.0000	0.0000	0.0000

Appendix B: Table of the Standard Gaussian Distribution

Standard Gaussian distribution $F(x) = \text{Prob}[Z \leq x]$

x	$F(x)$	x	$F(x)$	x	$F(x)$	x	$F(x)$
0.00	0.5000	0.50	0.6915	1.00	0.8413	1.50	0.9332
0.01	0.5040	0.51	0.6950	1.01	0.8438	1.51	0.9345
0.02	0.5080	0.52	0.6985	1.02	0.8461	1.52	0.9357
0.03	0.5120	0.53	0.7019	1.03	0.8485	1.53	0.9370
0.04	0.5160	0.54	0.7054	1.04	0.8508	1.54	0.9382
0.05	0.5199	0.55	0.7088	1.05	0.8531	1.55	0.9394
0.06	0.5239	0.56	0.7123	1.06	0.8554	1.56	0.9406
0.07	0.5279	0.57	0.7157	1.07	0.8577	1.57	0.9418
0.08	0.5319	0.58	0.7190	1.08	0.8599	1.58	0.9429
0.09	0.5359	0.59	0.7224	1.09	0.8621	1.59	0.9441
0.10	0.5398	0.60	0.7257	1.10	0.8643	1.60	0.9452
0.11	0.5438	0.61	0.7291	1.11	0.8665	1.61	0.9463
0.12	0.5478	0.62	0.7324	1.12	0.8686	1.62	0.9474
0.13	0.5517	0.63	0.7357	1.13	0.8708	1.63	0.9484
0.14	0.5557	0.64	0.7389	1.14	0.8729	1.64	0.9495
0.15	0.5596	0.65	0.7422	1.15	0.8749	1.65	0.9505
1.16	0.5636	0.66	0.7454	1.16	0.8770	1.66	0.9515
0.17	0.5675	0.67	0.7486	1.17	0.8790	1.67	0.9525
0.18	0.5714	0.68	0.7517	1.18	0.8810	1.68	0.9535
0.19	0.5753	0.69	0.7549	1.19	0.8830	1.69	0.9545
0.20	0.5793	0.70	0.7580	1.20	0.8849	1.70	0.9554
0.21	0.5832	0.71	0.7611	1.21	0.8869	1.71	0.9564
0.22	0.5871	0.72	0.7642	1.22	0.8888	1.72	0.9573
0.23	0.5910	0.73	0.7673	1.23	0.8907	1.73	0.9582
0.24	0.5948	0.74	0.7704	1.24	0.8925	1.74	0.9591
0.25	0.5987	0.75	0.7734	1.25	0.8944	1.75	0.9599
0.26	0.6026	0.76	0.7764	1.26	0.8962	1.76	0.9608
0.27	0.6064	0.77	0.7794	1.27	0.8980	1.77	0.9616
0.28	0.6103	0.78	0.7823	1.28	0.8997	1.78	0.9625
0.29	0.6141	0.79	0.7852	1.29	0.9015	1.79	0.9633
0.30	0.6179	0.80	0.7881	1.30	0.9032	1.80	0.9641
0.31	0.6217	0.81	0.7910	1.31	0.9049	1.81	0.9649
0.32	0.6255	0.82	0.7939	1.32	0.9066	1.82	0.9656
0.33	0.6293	0.83	0.7967	1.33	0.9082	1.83	0.9664
0.34	0.6331	0.84	0.7995	1.34	0.9099	1.84	0.9671

Source: *Handbook of Tables for Probability and Statistics*, 2d ed., W. H. Beyer, ed. (West Palm Beach, Fla.: CRC Press, Inc., 1968). Adapted with permission of the publisher.

Standard Gaussian distribution $F(x) = \text{Prob}[Z \leq x]$

x	$F(x)$	x	$F(x)$	x	$F(x)$	x	$F(x)$
0.35	0.6368	0.85	0.8023	1.35	0.9115	1.85	0.9678
0.36	0.6406	0.86	0.8051	1.36	0.9131	1.86	0.9686
0.37	0.6443	0.87	0.8078	1.37	0.9147	1.87	0.9693
0.38	0.6480	0.88	0.8106	1.38	0.9162	1.88	0.9699
0.39	0.6517	0.89	0.8133	1.39	0.9177	1.89	0.9706
0.40	0.6554	0.90	0.8159	1.40	0.9192	1.90	0.9713
0.41	0.6591	0.91	0.8186	1.41	0.9207	1.91	0.9719
0.42	0.6628	0.92	0.8212	1.42	0.9222	1.92	0.9726
0.43	0.6664	0.93	0.8238	1.43	0.9236	1.93	0.9732
0.44	0.6700	0.94	0.8264	1.44	0.9251	1.94	0.9738
0.45	0.6736	0.95	0.8289	1.45	0.9265	1.95	0.9744
0.46	0.6772	0.96	0.8315	1.46	0.9279	1.96	0.9750
0.47	0.6808	0.97	0.8340	1.47	0.9292	1.97	0.9756
0.48	0.6844	0.98	0.8365	1.48	0.9306	1.98	0.9761
0.49	0.6879	0.99	0.8389	1.49	0.9319	1.99	0.9767
0.50	0.6915	1.00	0.8413	1.50	0.9332	2.00	0.9772
2.00	0.9773	2.50	0.9938	3.00	0.9987	3.50	0.9998
2.01	0.9778	2.51	0.9940	3.01	0.9987	3.51	0.9998
2.02	0.9783	2.52	0.9941	3.02	0.9987	3.52	0.9998
2.03	0.9788	2.53	0.9943	3.03	0.9988	3.53	0.9998
2.04	0.9793	2.54	0.9945	3.04	0.9988	3.54	0.9998
2.05	0.9798	2.55	0.9946	3.05	0.9989	3.55	0.9998
2.06	0.9803	2.56	0.9948	3.06	0.9989	3.56	0.9998
2.07	0.9808	2.57	0.9949	3.07	0.9989	3.57	0.9998
2.08	0.9812	2.58	0.9951	3.08	0.9990	3.58	0.9998
2.09	0.9817	2.59	0.9952	3.09	0.9990	3.59	0.9998
2.10	0.9821	2.60	0.9953	3.10	0.9990	3.60	0.9998
2.11	0.9826	2.61	0.9955	3.11	0.9991	3.61	0.9998
2.12	0.9830	2.62	0.9956	3.12	0.9991	3.62	0.9999
2.13	0.9834	2.63	0.9957	3.13	0.9991	3.63	0.9999
2.14	0.9838	2.64	0.9959	3.14	0.9992	3.64	0.9999
2.15	0.9842	2.65	0.9960	3.15	0.9992	3.65	0.9999
2.16	0.9846	2.66	0.9961	2.16	0.9992	3.66	0.9999
2.17	0.9850	2.67	0.9962	3.17	0.9992	3.67	0.9999
2.18	0.9854	2.68	0.9963	3.18	0.9993	3.68	0.9999
2.19	0.9857	2.69	0.9964	3.19	0.9993	3.69	0.9999
2.20	0.9861	2.70	0.9965	3.20	0.9993	3.70	0.9999
2.21	0.9864	2.71	0.9966	3.21	0.9993	3.71	0.9999
2.22	0.9868	2.72	0.9967	3.22	0.9994	3.72	0.9999
2.23	0.9871	2.73	0.9968	2.23	0.9994	3.73	0.9999
2.24	0.9875	2.74	0.9969	3.24	0.9994	3.74	0.9999

Standard Gaussian distribution $F(x) = \mathrm{Prob}[Z \leq x]$

x	$F(x)$	x	$F(x)$	x	$F(x)$	x	$F(x)$
2.25	0.9878	2.75	0.9970	3.25	0.9994	3.75	0.9999
2.26	0.9881	2.76	0.9971	3.26	0.9994	3.76	0.9999
2.27	0.9884	2.77	0.9972	3.27	0.9995	3.77	0.9999
2.28	0.9887	2.78	0.9973	3.28	0.9995	3.78	0.9999
2.29	0.9890	2.79	0.9974	3.29	0.9995	3.79	0.9999
2.30	0.9893	2.80	0.9974	3.30	0.9995	3.80	0.9999
2.31	0.9896	2.81	0.9975	3.31	0.9995	3.81	0.9999
2.32	0.9898	2.82	0.9976	3.32	0.9995	3.82	0.9999
2.33	0.9901	2.83	0.9977	3.33	0.9996	3.83	0.9999
2.34	0.9904	2.84	0.9977	3.34	0.9996	3.84	0.9999
2.35	0.9906	2.85	0.9978	3.35	0.9996	3.85	0.9999
2.36	0.9909	2.86	0.9979	3.36	0.9996	3.86	0.9999
2.37	0.9911	2.87	0.9979	3.37	0.9996	3.87	0.9999
2.38	0.9913	2.88	0.9980	3.38	0.9996	3.88	0.9999
2.39	0.9916	2.89	0.9981	3.39	0.9997	3.89	1.0000
2.40	0.9918	2.90	0.9981	3.40	0.9997	3.90	1.0000
2.41	0.9920	2.91	0.9982	3.41	0.9997	3.91	1.0000
2.42	0.9922	2.92	0.9982	3.42	0.9997	3.92	1.0000
2.43	0.9925	2.93	0.9983	3.43	0.9997	3.93	1.0000
2.44	0.9927	2.94	0.9984	3.44	0.9997	3.94	1.0000
2.45	0.9929	2.95	0.9984	3.45	0.9997	3.95	1.0000
2.46	0.9931	2.96	0.9985	3.46	0.9997	3.96	1.0000
2.47	0.9932	2.97	0.9985	3.47	0.9997	3.97	1.0000
2.48	0.9934	2.98	0.9986	3.48	0.9997	3.98	1.0000
2.49	0.9936	2.99	0.9986	3.49	0.9998	3.99	1.0000
2.50	0.9938	3.00	0.9987	3.50	0.9998	4.00	1.0000

Appendix C: Table of Percentiles of Student's *t* Distribution

Percentiles of Student's *t* distribution

Degrees of freedom \ Percentile	60	75	90	95	97.5	99	99.5	99.95
1	0.325	1.000	3.078	6.314	12.706	31.821	63.657	636.619
2	0.289	0.816	1.886	2.920	4.303	6.965	9.925	31.598
3	0.277	0.765	1.638	2.353	3.182	4.541	5.841	12.924
4	0.271	0.741	1.533	2.132	2.776	3.747	4.604	8.610
5	0.267	0.727	1.476	2.015	2.571	3.365	4.032	6.869
6	0.265	0.718	1.440	1.943	2.447	3.143	3.707	5.959
7	0.263	0.711	1.415	1.895	2.365	2.998	3.499	5.408
8	0.262	0.706	1.397	1.860	2.306	2.896	3.355	5.041
9	0.261	0.703	1.383	1.833	2.262	2.821	3.250	4.781
10	0.260	0.700	1.372	1.812	2.228	2.764	3.169	4.587
11	0.260	0.697	1.363	1.796	2.201	2.718	3.106	4.437
12	0.259	0.695	1.356	1.782	2.179	2.681	3.055	4.318
13	0.259	0.694	1.350	1.771	2.160	2.650	3.012	4.221
14	0.258	0.692	1.345	1.761	2.145	2.624	2.977	4.140
15	0.258	0.691	1.341	1.753	2.131	2.602	2.947	4.073
16	0.258	0.690	1.337	1.746	2.120	2.583	2.921	4.015
17	0.257	0.689	1.333	1.740	2.110	2.567	2.898	3.965
18	0.257	0.688	1.330	1.734	2.101	2.552	2.878	3.922
19	0.257	0.688	1.328	1.729	2.093	2.539	2.861	3.883
20	0.257	0.687	1.325	1.725	2.086	2.528	2.845	3.850
21	0.257	0.686	1.323	1.721	2.080	2.518	2.831	3.819
22	0.256	0.686	1.321	1.717	2.074	2.508	2.819	3.792
23	0.256	0.685	1.319	1.714	2.069	2.500	2.807	3.767
24	0.256	0.685	1.318	1.711	2.064	2.492	2.797	3.745
25	0.256	0.684	1.316	1.708	2.060	2.485	2.787	3.725
26	0.256	0.684	1.315	1.706	2.056	2.479	2.779	3.707
27	0.256	0.684	1.314	1.703	2.052	2.473	2.771	3.690
28	0.256	0.683	1.313	1.701	2.048	2.467	2.763	3.674
29	0.256	0.683	1.311	1.699	2.045	2.462	2.756	3.659
30	0.256	0.683	1.310	1.697	2.042	2.457	2.750	3.646
40	0.255	0.681	1.303	1.684	2.021	2.423	2.704	3.551
60	0.254	0.679	1.296	1.671	2.000	2.390	2.660	3.460
120	0.254	0.677	1.289	1.658	1.980	2.358	2.617	3.373
∞	0.253	0.674	1.282	1.645	1.960	2.326	2.576	3.291

Source: *Handbook of Tables for Probability and Statistics*, 2d ed., W. H. Beyer, ed. (West Palm Beach, Fla.: CRC Press, Inc., 1968). Adapted with permission of the publisher.

Appendix D: Table of 95th Percentiles of the F Distribution

95th percentiles of the F distribution

$n-k-1$ \ k	1	2	3	4	5	6	7	8	9
1	161.4	199.5	215.7	224.6	230.2	234.0	236.8	238.9	240.5
2	18.51	19.00	19.16	19.25	19.30	19.33	19.35	19.37	19.38
3	10.13	9.55	9.28	9.12	9.01	8.94	8.89	8.85	8.81
4	7.71	6.94	6.59	6.39	6.26	6.16	6.09	6.04	6.00
5	6.61	5.79	5.41	5.19	5.05	4.95	4.88	4.82	4.77
6	5.99	5.14	4.76	4.53	4.39	4.28	4.21	4.15	4.10
7	5.59	4.74	4.35	4.12	3.97	3.87	3.79	3.73	3.68
8	5.32	4.46	4.07	3.84	3.69	3.58	3.50	3.44	3.39
9	5.12	4.26	3.86	3.63	3.48	3.37	3.29	3.23	3.18
10	4.96	4.10	3.71	3.48	3.33	3.22	3.14	3.07	3.02
11	4.84	3.98	3.59	3.36	3.20	3.09	3.01	2.95	2.90
12	4.75	3.89	3.49	3.26	3.11	3.00	2.91	2.85	2.80
13	4.67	3.81	3.41	3.18	3.03	2.92	2.83	2.77	2.71
14	4.60	3.74	3.34	3.11	2.96	2.85	2.76	2.70	2.65
15	4.54	3.68	3.29	3.06	2.90	2.79	2.71	2.64	2.59
16	4.49	3.63	3.24	3.01	2.85	2.74	2.66	2.59	2.54
17	4.45	3.59	3.20	2.96	2.81	2.70	2.61	2.55	2.49
18	4.41	3.55	3.16	2.93	2.77	2.66	2.58	2.51	2.46
19	4.38	3.52	3.13	2.90	2.74	2.63	2.54	2.48	2.42
20	4.35	3.49	3.10	2.87	2.71	2.60	2.51	2.45	2.39
21	4.32	3.47	3.07	2.84	2.68	2.57	2.49	2.42	2.37
22	4.30	3.44	3.05	2.82	2.66	2.55	2.46	2.40	2.34
23	4.28	3.42	3.03	2.80	2.64	2.53	2.44	2.37	2.32
24	4.26	3.40	3.01	2.78	2.62	2.51	2.42	2.36	2.30
25	4.24	3.39	2.99	2.76	2.60	2.49	2.40	2.34	2.28
26	4.23	3.37	2.98	2.74	2.59	2.47	2.39	2.32	2.27
27	4.21	3.35	2.96	2.73	2.57	2.46	2.37	2.31	2.25
28	4.20	3.34	2.95	2.71	2.56	2.45	2.36	2.29	2.24
29	4.18	3.33	2.93	2.70	2.55	2.43	2.35	2.28	2.22
30	4.17	3.32	2.92	2.69	2.53	2.42	2.33	2.27	2.21
40	4.08	3.23	2.84	2.61	2.45	2.34	2.25	2.18	2.12
60	4.00	3.15	2.76	2.53	2.37	2.25	2.17	2.10	2.04
120	3.92	3.07	2.68	2.45	2.29	2.17	2.09	2.02	1.96
∞	3.84	3.00	2.60	2.37	2.21	2.10	2.01	1.94	1.88

Source: *Handbook of Tables for Probability and Statistics*, 2d ed., W. H. Beyer, ed. (West Palm Beach, Fla.: CRC Press, Inc., 1968). Adapted with permission of the publisher.
Note: k = number of predictor variables; n = number of data points.

10	12	15	20	24	30	40	60	120	∞
241.9	243.9	245.9	248.0	249.1	250.1	251.1	252.2	253.3	254.3
19.40	19.41	19.43	19.45	19.45	19.46	19.47	19.48	19.49	19.50
8.79	8.74	8.70	8.66	8.64	8.62	8.59	8.57	8.55	8.53
5.96	5.91	5.86	5.80	5.77	5.75	5.72	5.69	5.66	5.63
4.74	4.68	4.62	4.56	4.53	4.50	4.40	4.43	4.40	4.36
4.06	4.00	3.94	3.87	3.84	3.81	3.77	3.74	3.70	3.67
3.64	3.57	3.51	3.44	3.41	3.38	3.34	3.30	3.27	3.23
3.35	3.28	3.22	3.15	3.12	3.08	3.04	3.01	2.97	2.93
3.14	3.07	3.01	2.94	2.90	2.86	2.83	2.79	2.75	2.71
2.98	2.91	2.85	2.77	2.74	2.70	2.66	2.62	2.58	2.54
2.85	2.79	2.72	2.65	2.61	2.57	2.53	2.49	2.45	2.40
2.75	2.69	2.62	2.54	2.51	2.47	2.43	2.38	2.34	2.30
2.67	2.60	2.53	2.46	2.42	2.38	2.34	2.30	2.25	2.21
2.60	2.53	2.46	2.39	2.35	2.31	2.27	2.22	2.18	2.13
2.54	2.48	2.40	2.33	2.29	2.25	2.20	2.16	2.11	2.07
2.49	2.42	2.35	2.28	2.24	2.19	2.15	2.11	2.06	2.01
2.45	2.38	2.31	2.23	2.19	2.15	2.10	2.06	2.01	1.96
2.41	2.34	2.27	2.19	2.15	2.11	2.06	2.02	1.97	1.92
2.38	2.31	2.23	2.16	2.11	2.07	2.03	1.98	1.93	1.88
2.35	2.28	2.20	2.12	2.08	2.04	1.90	1.95	1.90	1.84
2.32	2.25	2.18	2.10	2.05	2.01	1.96	1.92	1.87	1.81
2.30	2.23	2.15	2.07	2.03	1.98	1.94	1.89	1.84	1.78
2.27	2.20	2.13	2.05	2.01	1.96	1.91	1.86	1.81	1.76
2.25	2.18	2.11	2.03	1.98	1.94	1.89	1.84	1.79	1.73
2.24	2.16	2.09	2.01	1.90	1.92	1.87	1.82	1.77	1.71
2.22	2.15	2.07	1.99	1.95	1.90	1.85	1.80	1.75	1.69
2.20	2.13	2.06	1.97	1.93	1.88	1.84	1.79	1.73	1.67
2.19	2.12	2.04	1.96	1.91	1.87	1.82	1.77	1.71	1.65
2.18	2.10	2.03	1.94	1.90	1.85	1.81	1.75	1.70	1.64
2.16	2.09	2.01	1.93	1.89	1.84	1.79	1.74	1.68	1.62
2.08	2.00	1.92	1.84	1.79	1.74	1.69	1.64	1.58	1.51
1.99	1.92	1.84	1.75	1.70	1.65	1.59	1.53	1.47	1.39
1.91	1.83	1.75	1.66	1.61	1.55	1.50	1.43	1.35	1.25
1.83	1.75	1.67	1.57	1.52	1.46	1.39	1.32	1.22	1.00

Problem Sources

1.1

J. P. Jarvis, K. Stevenson, and T. Willemain. "A Simple Procedure for Ambulance Allocation in Areas of Low Demand." In *Emergency Medical Systems Analysis*. Lexington, Mass.: Lexington Books, 1977.

1.2

U. S., Department of Commerce, Bureau of the Census. *Construction Reports*. Washington, D. C.: Government Printing Office, 1960–1974.

1.3

H. T. Taggart. *Home Mortgage Lending Patterns in Metropolitan Boston*. Commonwealth of Massachusetts Publication 10252-45-300-2-78-CR, Boston, December 1977.

2.1

Massachusetts Taxpayers Foundation, Inc. "Municipal Financial Data, Including 1977 Tax Rates." Boston, 1977.

2.3

T. Willemain. "Neighborhood Health Centers in Cambridge." Laboratory of Architecture and Planning, Massachusetts Institute of Technology, 1976.

3.1

K. McClure. "An Evaluation of the Rent Control Policy of Cambridge, Massachusetts." Masters thesis, Department of Urban Studies and Planning, Massachusetts Institute of Technology, 1978.

3.4

Massachusetts Taxpayers Foundation, Inc. "Municipal Financial Data, Including 1977 Tax Rates." Boston, 1977.

3.5

S. C. Hadaway. "Diversification Possibilities in Agricultural Land Investments." *Appraisal Journal* 44 (1978): 529–537. The opinions and statements set forth herein are those of the contributor(s) and do not necessarily reflect the viewpoint of the American Institute of Real Estate Appraisers or its individual members, and neither the Institute nor its editors and staff assume responsibility for such expressions of opinion or statements.

4.3

J. F. Schnelle, et al. "Patrol Evaluation Research: A Multiple Baseline Analysis of Saturation Police Patrolling during Day and Night Hours." *Journal of Applied Behavior Analysis* 10 (1977): 33–40. Copyright 1977 by the Society for the Experimental Analysis of Behavior, Inc.

4.4

J. J. Heckman and R. J. Willis. "A Beta-logistic Model for the Analysis of Sequential Labor Force Participation by Married Women." *Journal of Political Economy* 85 (1977): 27–58.

4.5

N. Geller and M. Yochmowitz. "Regional Planning of Maternity Services." *Health Services Research* (Spring 1975), p. 63.

4.6

U. S., Department of Commerce, Bureau of the Census. *Statistical Abstract of the United States 1977.* Washington, D. C.: Government Printing Office, 1977, p. 440.

5.2

E. Seligman. "Finance Statistics of the American Commonwealths." *Publications of the American Statistical Association* 1 (1889): 349–468.

6.1

N. Okonjo. "Indigenous Rural Savings and Credit Systems: A Case Study from Bendel State, Nigeria." Masters thesis, Department of Urban Studies and Planning, Massachusetts Institute of Technology, 1978.

6.2

N. Okonjo. "Indigenous Rural Savings and Credit Systems: A Case Study from Bendel State, Nigeria." Masters thesis, Department of Urban Studies and Planning, Massachusetts Institute of Technology, 1978.

6.4

L. Branch and F. Fowler, Jr. "The Health Care Needs of the Elderly and Chronically Disabled in Massachusetts." Center for Survey Research, Boston, March 1975, pg. 13.

Massachusetts, Department of Public Health. "Report of the Long Term Care Task Force." Boston, August 1977, pp. 9,14.

8.2

H. Taggart. *Home Mortgage Lending Patterns in Metropolitan Boston.* Commonwealth of Massachusetts Publication 10252-45-300-2-78-CR, Boston, December 1977.

8.4

N. Okonjo. "Indigenous Rural Savings and Credit Systems: A Case Study from Bendel State, Nigeria." Masters thesis, Department of Urban Studies and Planning, Massachusetts Institute of Technology, 1978.

9.2

U. S., Department of Commerce, Bureau of the Census. *County and City Data Book.* Washington, D. C.: Government Printing Office, 1977.

9.3

U. S., Department of Housing and Urban Development, Community Development Block Grant Program. *Directory of Allocations for Fiscal Year 1975.* Washington, D. C.: Government Printing Office, 1975, p. 28.

9.4

S. C. Hadaway. "Diversification Possibilities in Agricultural Land Investments." *Appraisal Journal* 44 (1978): 529–537. The opinions and statements set forth herein are those of the contributor(s) and do not necessarily reflect the viewpoint of the American Institute of Real Estate Appraisers or its individual members, and neither the Institute nor its editors and staff assume responsibility for such expressions of opinion or statements.

11.1

T. Willemain, J. Coldiron, K. Natwin, and D. Frost. Unpublished data.

11.3

Mobile Homes Manufacturers Association. In U. S., Department of Housing and Urban Development, *1976 Statistical Yearbook*. Washington, D. C.: Government Printing Office, 1977, p. 283.

Index

Statistical Methods for Planners